OUR STORY YOUR HISTORY
THE INTERNATIONAL
BOMBER COMMAND CENTRE

OUR STORY YOUR HISTORY

THE INTERNATIONAL BOMBER COMMAND CENTRE

STEVE DARLOW
MARK DODDS
DAN ELLIN
SEAN FEAST
ROBERT OWEN

Published in 2018 by Fighting High Ltd
www.fightinghigh.com

British Library Cataloguing-in-Publication data. A CIP record for this title is available from the British Library.

ISBN – 13-978-1-9998128-2-9

IBCC and Bomber Command veteran photography by Paul Mellor.
www.paulmphoto.co.uk

Jacket and Book Designed by TruthStudio Limited.
www.truthstudio.co.uk

Printed and bound in Wales by Gomer Press.

In memory of Tony Worth CVO.

CONTENTS

THOMSON HD • THOMSON HJ
J 064 • THOMSON J 186 • THOMSON J 465
JA 228 • THOMSON JH • THOMSON JS
R • THOMSON RC • THOMSON RH • THOMSON TA
W 347 • THOMSON WB • THOMSON WH
J • THORN CM • THORN ES • THORN RL
CJ • THORNE JW • THORNE P • THORNE WJ
THORNHILL W • THORNHILL WH • THORNLEY GH
ER • THORNTON FC • THORNTON FW
JH 977 •
JH 631 • THORNTON KF
REJ •
W • THORNTON WC
R • THOROLDSON WE
HJ •
PE DM •
PG • THORPE RHA
THROWER CE
THRASHER JW
HG •
EE • THURSTON
THYRET HR •
HJR • TIDBY CH •
CG • TILBROOK JM •
TILLARD JN • TILLEN FI •
TILSON CG • TILTON AA •
TIMPERLEY LJ •
DW • TIMMS L •
TINDALE VJS • TINDELL DW
GC • TINKLER AE • TINSLEY
TINKER J •
TIPPLE AR •
TOOMEY HA • TOOMEY L • TOON JW • TOOTH A •
TOPHAM G • TOPHAM LK • TOPHAM S • TOPLIS JT •
TORBETT VWG • TORKELSON DL • TORNEY T •
TOSH JF • TOTHILL FHN • TOTTEN J •
TOUGH GH • TOUGH JGR • TOULSON R • TOUPIN JOT •
TOVERY F • TOVEY KJ • TOWERS AF •
TOWLE EG • TOWLE JG •
TOWNDROW R •
TOWNSEND AE •
TOWNSEND ET
TOWNSEND R •
TRAQUAIR WLE

FOREWORD

Left Squadron Leader George 'Johnny' Johnson, MBE, DFM, initially flew on operations as an air gunner with No. 97 Squadron prior to training as a specialist bomb aimer. Following completion of his operational tour and subsequent training duties he then became part of No. 617 Squadron, taking part in the Dam Buster raid of 16/17 May 1943, attacking the Sorpe dam. In 2017 Johnny was awarded an honorary doctorate by the University of Lincoln for his contribution to British society, and in November that year Her Majesty Queen Elizabeth II presented Johnny with his MBE at Buckingham Palace.

When, three years ago, I first became aware of the plan to build a Bomber Command memorial in Lincoln, I was immediately struck by the imagination and vision of this project.

The International Bomber Command Centre is a unique undertaking. Here, in one place, is a fitting memorial to almost 58,000 young men and women, from all over the world, who fought in Bomber Command between 1939 and 1945 and died in the service of our country, protecting our way of life. As important as honouring those who passed away, is the education centre that serves to demonstrate the role that Bomber Command played during the conflict, for the information of current and future generations – its history, what was done and how it was done. Through a series of interactive activities, young people in particular can come to understand why Bomber Command was so essential.

The siting is very noteworthy. Lincoln was at the very heart of Bomber Command during the war years. The memorial and centre atop Canwick Hill looks directly across the city valley to Lincoln Cathedral, a building that had such special significance for us in those days. On returning from missions we would rejoice at the sight of the cathedral towers in the early morning light. We were home.

I am one of the lucky ones who survived and consider myself very fortunate to be able to see the culmination of this wonderful project.

Squadron Leader George 'Johnny' Johnson, MBE, DFM,
Nos 97 and 617 Squadrons, 1942–44.

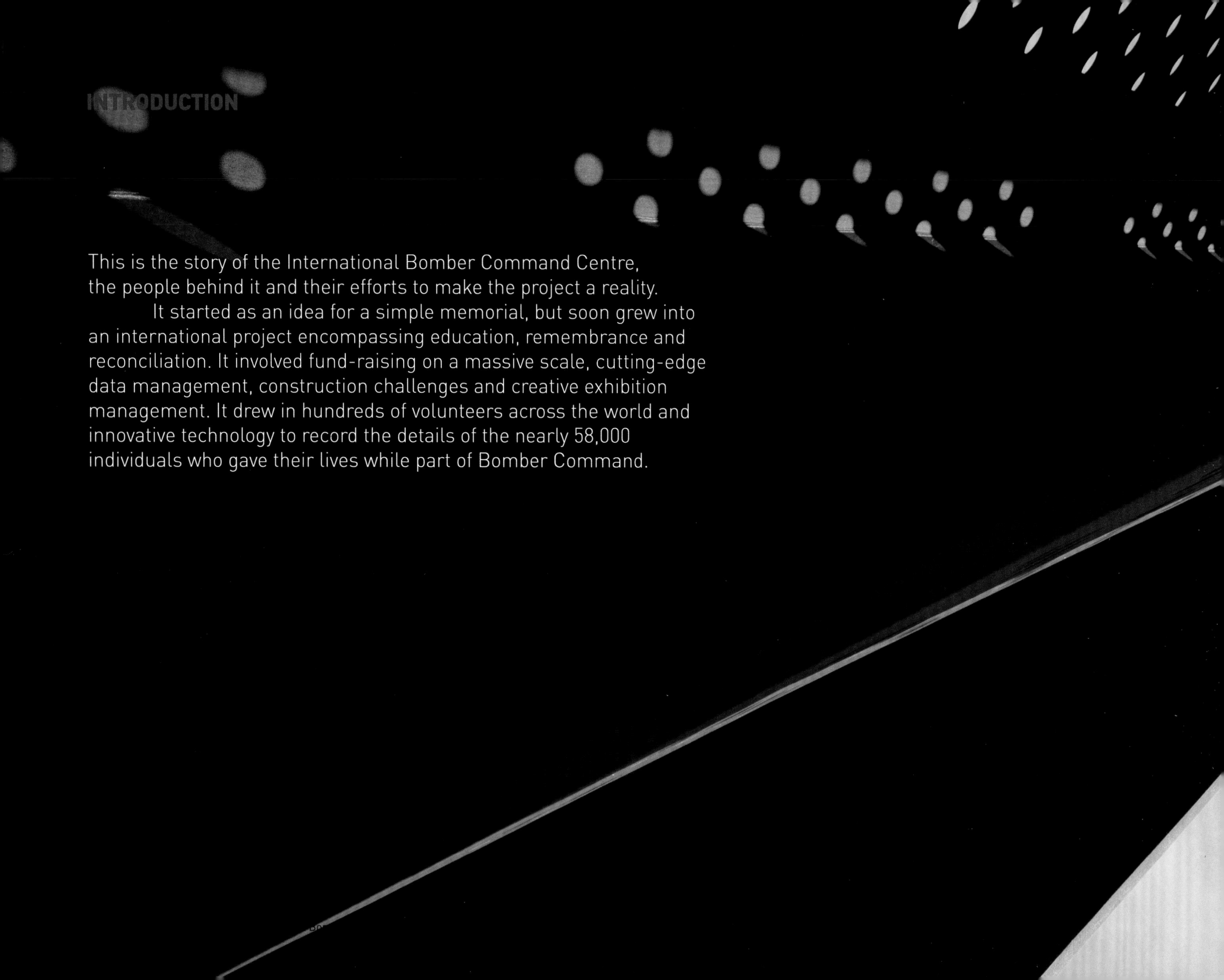

INTRODUCTION

This is the story of the International Bomber Command Centre, the people behind it and their efforts to make the project a reality.

It started as an idea for a simple memorial, but soon grew into an international project encompassing education, remembrance and reconciliation. It involved fund-raising on a massive scale, cutting-edge data management, construction challenges and creative exhibition management. It drew in hundreds of volunteers across the world and innovative technology to record the details of the nearly 58,000 individuals who gave their lives while part of Bomber Command.

THE INTERNATIONAL BOMBER COMMAND CENTRE
PART ONE ORIGINS

Left The IBCC Spire, set high on Canwick Hill, overlooking Lincoln, the city and Cathedral, in the bomber county of Lincolnshire.

'You think of these things and usually people are not interested, but it seems with this one they were.'

With these words from Tony Worth, the story of the International Bomber Command Centre began.

In early 2005 Tony Worth found himself in the north transept of Lincoln Cathedral. Like many such buildings, the cathedral has chapels dedicated to the armed forces and it was here that he noticed the Ledger stone, which was, according to the inscription: 'Dedicated to the men and women of Bomber Command over 55,000 of whom lost their lives.' The stone was positioned outside the door to the Royal Air Force chapel, a route sadly underused as a more natural route would take visitors through another entrance and miss this worthy memorial.

Tony thought about the importance of Bomber Command to the Allied efforts in the Second World War and the undeniable part the county of Lincolnshire played in that victory. Combining this with the less than admirable way in which Bomber Command had been treated after the war, he felt there should be something more significant to mark the debt paid by the brave individuals who were denied a medal and a place in the victory parade.

He was clear from the start that any memorial should be just the start of the story of discovery. A memorial is just that – a monument to a group of people. Tony, however, felt the need to articulate the accounts of the people behind the numbers as each of them had a human story to tell. So rather than another museum full of 'bent pieces of metal' and aircraft, he wanted somewhere where the people could be remembered, from the ground crew and support teams to the pilots, bomb aimers and gunners. And so it was that the concept of the Bomber Command Centre was born.

His first task was to find a site for a memorial. Initially he found a suitable position on South Common in Lincoln, because it was common land and 'of the people', but it was quickly discovered that an Act of Parliament (the Lincoln City Council Act 1985) meant that nothing permanent could be built on the land, so an alternative had to be found.

Undeterred, Tony ventured a few hundred yards further south and identified a new site at the top of Canwick Hill, an imposing position with a sweeping view of Lincoln, the cathedral and the surrounding countryside. Further research showed that the land in question belonged to Jesus College Oxford, so Tony spoke to the college, who after discussions agreed to rent the ten acres at an agricultural rent for 125 years. This gave the Centre its first physical base and the opportunity to develop the site with a welcome and assured long-term future.

And so the mission had begun and many of the project team would readily admit that they did not know the huge and significant scale it would take on, nor how it would gain such a positive momentum and develop into a truly international operation.

Realising that support from a wide range of people would be critical to the venture, in 2009 Tony contacted Air Marshal Sir Michael Beetham, the then President of the Bomber Command Association. While agreeing to the concept, Sir Michael asked Tony to 'hold fire' on the public side of the scheme as work was already under way on the London Bomber Command Memorial, sited in The Green Park. Of course, Tony generously offered the Lincoln site as an alternative, but it was felt that a memorial in the capital would be more suitable. Agreeing to hold off any fund-raising until the London memorial was built and paid for, Tony and his now small team started their 'homework', and managed to apply successfully for a small grant from the Arts Council of England to produce the memorial designs.

With the funding from the Arts Council secured, a competition was launched to design the Lincolnshire

Left Tony Worth CVO, whose vision inspired the International Bomber Command Centre project.

Below Tony Worth and Bomber Command veteran Charles Clarke at the ground breaking ceremony for the Chadwick Centre, October 2016.

Bomber Command memorial. Many impressive entries were received, and they were eventually whittled down by committee to a shortlist of three. These were voted on by a wide audience including veterans, members of the local community and those closely involved with the project. The winning design, a Memorial Spire, was chosen and the project was formally launched on 31 May 2013 (coincidentally the anniversary of the first 'thousand-bomber' raid on Cologne).

The original design for the Spire was that it should be some fifty metres high, with the names of those who lost their lives making up the structure of the Spire itself. Further research, however, showed that there could be technical difficulties with achieving this result, not the least of which was that it was felt that families may find it difficult to locate their relations' names on such a tall structure. In addition, English Heritage advised that the structure as originally designed would be too high and detract from the view of Lincoln Cathedral. As a solution, it was suggested that the Spire height could be amended to, rather fittingly, 102 feet, the same as the wingspan of the Lancaster bomber that was such an integral part of the history of Bomber Command.

With the names no longer being part of the structure of the Spire itself, the team needed to identify a solution to remember the individuals – the cornerstone of the whole project. It was then that they introduced the concept of the memorial walls. Research showed that relatives not only wanted to see the names of their loved ones, but also to be able to touch them and perhaps attach a poppy. If the names were on the Spire this would not be possible, so the walls were conceived to carry the names of the fallen, and the number of poppies there every day are testament to how wise this decision was.

With this new concept in place, a local architect was commissioned to modify the plans to their current form, with the Spire based on the design and wingspan of the Lancaster, and the names of the individuals who gave their lives engraved on curved walls surrounding the Spire.

A charity was formed to give the project some structure, and in 2009 a Board of Trustees were selected to oversee the strategy and progress of the scheme. The trustees were chosen because of their extensive experience in project management for both private and public sectors, their networking abilities and specialist skills such as financial management. In 2013 a management board was established to oversee the delivery of the strategy and deal with operational issues. At this stage the team was very small, with just two members, but in 2015 this increased to eight. More specialist panels were then developed to look at other aspects of the venture: the Programme (digital archive, exhibition design, outreach and engagement); Development (marketing, fund-raising, PR, construction and project development); and People (budgetary management, staff and volunteer recruitment, compliance and health and safety).

As a work to preserve history, it was quite simply vital, and with the 'corporate' structure in place, the vision of the project became clear. It was to become a centre for recognition, remembrance, education and reconciliation.

But with the core aim of recording the memories of those who had served, it was quite literally a race against time. The veterans were not getting any younger and action had to be taken to make sure their memories were recorded before it was simply too late.

Purpose, Location And Scope

The International Bomber Command Centre is an immense project. From small beginnings it has evolved into a global centre, leading the world as an information resource for those who served and lost their lives with Bomber Command.

Why is the Centre located in Lincolnshire? Much of the reason is due to geography and landscape. Lincolnshire has many advantages for the location of airfields, with a relatively level landscape and its proximity to mainland Europe. In the First World War the county was home to thirty-eight landing grounds for the Royal Naval Air Service and the Royal Flying Corps. At the end of this war, many of these airstrips were given back over to the agricultural production for which Lincolnshire remains so well known today.

As the Second World War loomed and the threat of invasion was ever present, the nation began the counter-offensive – taking the fight to the enemy – and Bomber Command were tasked with playing a major part in this action. With Lincolnshire strategically located in the east of the country and offering prime land with good drainage and open skies, it became the natural region in which to position airfields for the heavy bombers that would prove

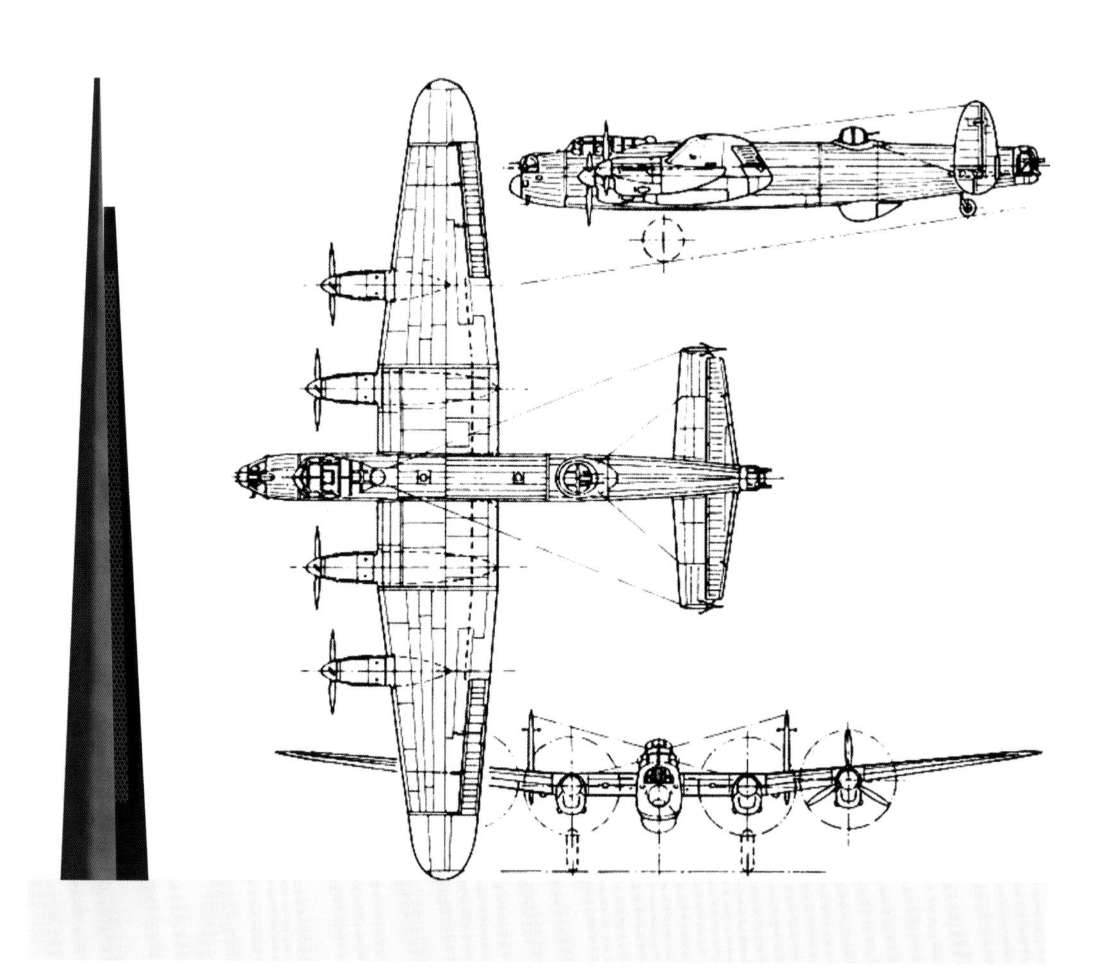

Left The height of the Memorial Spire matches the wingspan of the Avro Lancaster.

Below Design plans for the IBCC.

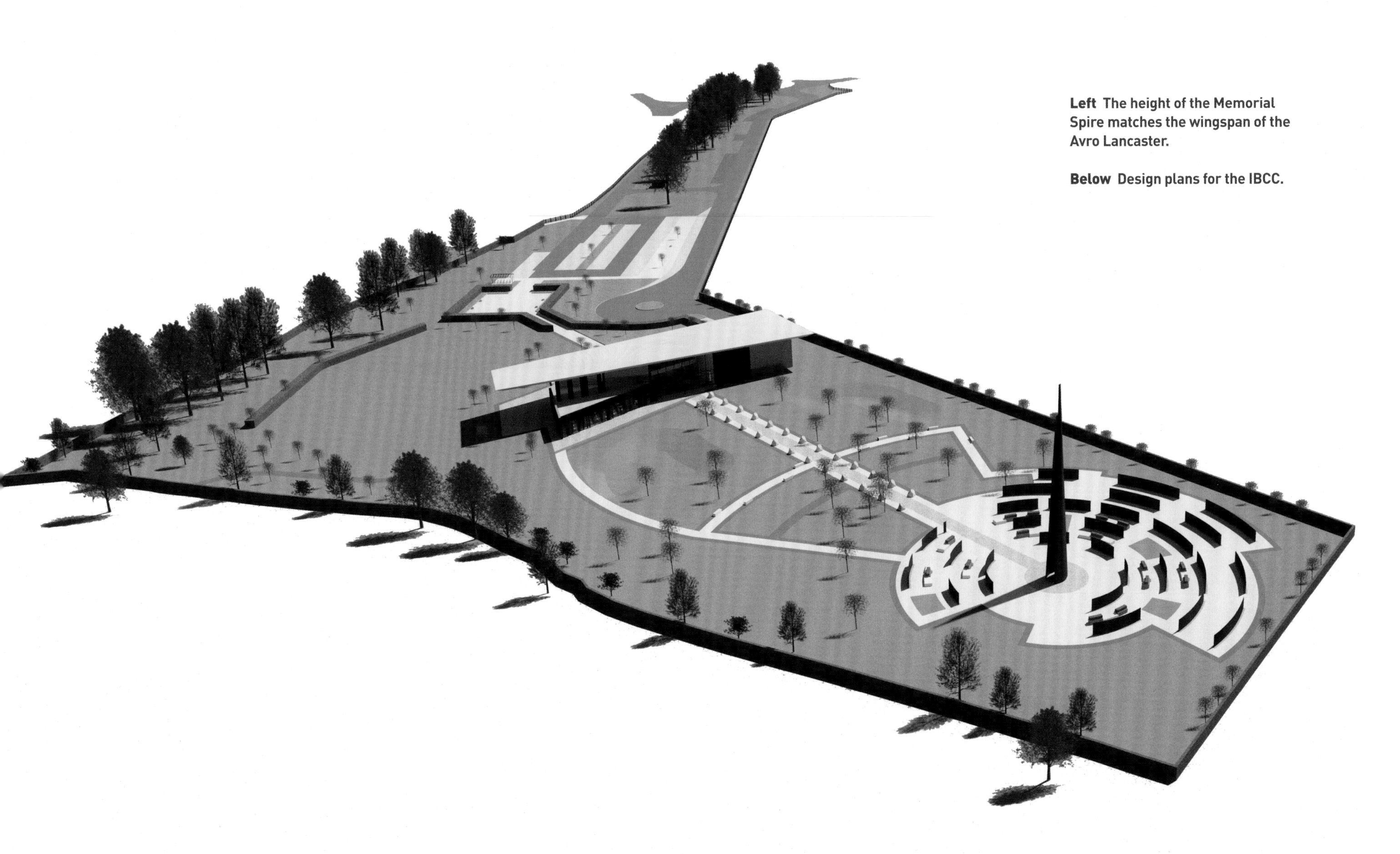

to be so important in the Allied war effort.

New airfields sprang up across the county. Paved runways and the all-important support roads and services were built, until at the end of the war there were in excess of forty airfields – twenty-seven of which were Bomber Command ones (a third of all the UK's stations) – giving Lincolnshire the proud mantle 'Bomber County'.

It was fully fitting, therefore, that a centre to remember the brave individuals who gave their lives for their nation was to be positioned in the very county in which many of them worked and took off from for the last time. And so it was that the Lincolnshire Bomber Command Memorial was conceived.

The increasing size and depth of the project generated further complexities. For example, once it was decided that the Centre would hold oral histories, there was a need to record them. It was decided to recruit volunteers to make the recordings and as they needed to be trained (wherever they were in the world), programmes were developed online to ensure consistency of message. Similarly, with one of the key aims of the venture being education, an exhibition centre was required, which brought in a whole new aspect in terms of physical construction and supporting finance.

But why was it felt that a central source of information on the Bomber Command losses was necessary? During their initial research, the project team found many websites where elements of information could be extracted, but the information was fragmented. It was often difficult to find and there just wasn't a reliable and authoritative source of information on the Bomber Command losses and the roles the individuals played in the Second World War.

The team considered building an archive of physical information, but this in itself raised a number of issues around storage, security, insurance and access, and in fact there were already many places such as the Imperial War Museum where physical archives existed. They felt it would be pointless moving objects from one place to another, so decided to take the digital route, which also offered many advantages in terms of access from across the globe, data management and data integrity. As they were proposing to use relatively new technology, the team realised that specialist advice was essential, and the University of Lincoln was brought on board because of their particular skills and experience in the practical issues involved with digital archiving and managing large volumes of interconnected data.

Additionally, it was decided that rather than simply 'ingest' the data that was already available in the data-bases such as those hosted by the Imperial War Museum, the Centre would focus on the information that was not immediately accessible – the voices of veterans telling their personal stories, as well as never-before-seen material from families that had often spent many years in bookcases and boxes under the bed.

Funding and Development

From the outset, the team were determined to make sure the Centre was viable and resilient. They also knew that it was unlikely to receive long-term Government funding, so they had to have the ability to make it work from the beginning.

The Heritage Lottery Fund (HLF) was approached for financial assistance. The initial stage of the application involved an Expression of Interest, which alone meant a massive effort in order to produce the detailed proposal of some 10,000 words. Once that was approved, the 'development bid' was produced. This again involved a tremendous amount of work as the team followed a long and very detailed application process, involving commissioning reports on audience development, exhibition design, and educational and activity programmes. This took just over a year and, of course, all cost money to produce. The HLF part-funded some of the work but around forty per cent of the expenses of the proposal had to be raised by the team.

The HLF bid process raised some interesting facts – one being that the team had underestimated the likely footfall, partly because in that time the project had grown from being Lincolnshire focused to being a truly international project. A rise in public interest had also played its part as news of the Centre and what it was trying to achieve spread across the world. Of course, if expected visitor numbers go up, so does the size and scope of facilities you need to handle them, such as the size of the exhibition centre, the cafe, number of toilets etc., so the team constantly re-thought the plans for what became known as the Chadwick Centre, named in honour of Roy Chadwick, designer of the Lancaster bomber, to incorporate the expanded facilities, while keeping an eye on the costs.

Numerous presentations and reports for the HLF followed. In many ways, for the HLF, it was an original project, as they were more used to looking at proposals in relation to buildings that already existed rather than a 'new build'. The proposal, therefore, was for the heritage aspects of the project, that is what was to be inside the Chadwick Centre and not the building of the Centre itself.

With the bid successful, the work began in earnest as the funding was only released retrospectively, to ensure the security of the funds, and then the HLF only paid a proportion of the whole costs. So rather than being a 'blank cheque' the HLF funding provided an incredibly valuable source of finance to keep the venture on track and help develop the Centre. This meant, of course, that once the funding was announced the team still had to continue their efforts to raise funds from elsewhere in order to keep the financial stream and the project alive.

Raising the Profile

Even before the International Bomber Command Centre opened, the team worked tirelessly on the development of the site, fund-raising and toiling to keep the profile of the Centre high both nationally and internationally. In 2016, the team hosted over 8,000 people from 9 different countries on visits to the site, including the involvement of over 600 children in tours or events. They also attended national tourism shows to raise the profile of the Centre with tour operators and holiday companies, to ensure that the Centre would be included as one of the main tourism attractions of the county, and for visits to be added as part of Lincolnshire itineraries. Just as importantly teams of volunteers attended major air shows and events to promote the project, communicate its objectives and of course raise much needed funds.

International Connections, International Branding

When you talk about Bomber Command, most people perhaps know that there were Canadians and possibly Polish nationals involved, but less known about is the involvement of the sixty-two other nationalities including German Jews, Nigerians and Chinese – all with one common aim. While they were not all nation supporters, the fact that individuals from those nations took part needed to be remembered. Some surprising tales emerged, including the Brazilians who chartered their own boats to come over and volunteer.

These international connections have been helped immensely by the project website and social media feeds, which have had engagement from over sixty-three countries. As the international aspects grew, it became more and more clear that there was a need to cover the whole of Bomber Command, and if this was to be done properly and have global appeal, the project needed to evolve from the originally intended Lincolnshire Bomber Command Memorial into the 'International Bomber Command Centre'.

A rebranding exercise was planned and when the new design for the building was complete the press were invited to the launch, which emphasised why the change was being made and the significance of the new international outlook.

Engagement

Engaging with the press was so important to the project, but it was difficult at times, primarily because the work of the Centre was not 'news'. Raising funds is a very competitive market and with thousands of good causes battling for space the team worked tirelessly to engage with the regional and national press. Many publications ran features on the veterans and appeals, which raised valuable funds, others engaged less, but the team were always grateful for any profile they could receive.

The involvement of the community was also critical, and local public support was never so high as when during building a break-in occurred on the site. It's fair to say there was outrage and within days the items stolen had been fully replaced (and more) by generous individuals and businesses. In addition, hundreds of people have undertaken sponsored runs, walks and cycle rides, held events in their back gardens and bought branded merchandise and tickets to larger-scale events run by the IBCC team themselves.

Because of the international aspect the team explored funding from individuals and foundations across the globe – using personal connections, links with museums and professional contacts. It all took time and a great deal of hard work but the team were determined to make sure the project reached completion.

Fund-raising was not without its challenges. The team sometimes found that the political hesitancy and apprehension over what Bomber Command did in the war as part of their operational duties is still present, and this

Right Tony Worth at the turf cutting for the Spire on 21 August 2014.

Below A gathering of Bomber Command veterans at the unveiling of the Spire on 2 October 2015.

sometimes got in the way of the fund-raising. However, when people rather than politicians got involved, the team often found a different story. Highlighting the international aspect of reconciliation, some of the most interesting work was done by German air crash investigators, who collaborated with families to find aircraft that had been brought down, and even identified pilots who were flying the fighters that fired the critical shots.

A Race against Time

One of the biggest challenges was without doubt raising the money for the project, but overriding this is the 'race against time'. Many of the veterans who have played such a large part in supporting the project have become very close friends with the IBCC team, which makes their passing so much more poignant and made the need to complete the project so much more urgent.

Looking after the veterans is vitally important and the team realise they have an immense duty of care towards the individuals. This goes as far as making sure that if they, for example, go to a press interview, they are chaperoned, have food and drink supplied and are in receipt of other vitally important aspects of care. The true spirit of the veterans shone through when the project was delayed for six months due to funding flows. The team wrote to all the veterans and their families explaining the delay. The many responses from the veterans came back in one of three forms: 'Real shame but had to be done', 'I don't think I'll make it but at least I'll die knowing it will be done', and 'I'd better live a little bit longer then!'

FAILED TO RETURN

V-VICTOR IS MISSING

1

Flying Officer George Agar is twenty-three and within sight of a rest. His service number, 41240, shows he has been in the RAF since before the war, one of many hundreds to have been granted a short service commission as the armed forces built up their strength in preparation for the inevitable battle that lay ahead.

His crew for the night of 25 April 1941 is one of nine operating from No. 218 Squadron out of Marham. They include Wing Commander Herbert Kirkpatrick, a regular of the old school, who has just taken over command of the squadron and has chosen this as his first operation in charge.

Within Wellington R1507 HA-V 'Victor' with George at the controls is most of his regular crew: Flight Sergeant Clifford Andrews; Sergeant Victor Ashworth; and Flight Sergeant Wilfred Thornhill. All three were with their skipper when he completed his operation in a Wimpey on 9 January. Indeed, Thornhill has been with Agar since the late summer of 1940, when they flew their first sorties together in a Blenheim IV to attack enemy coastal targets and barge concentrations in Holland and France. While the fighter boys were doing their bit above the fields of Kent and beyond, the bomber boys were also doing their level best to prevent a German invasion.

One of the air gunners in the Wellington that night is a relative novice, but only to bomber operations. Pilot Officer Charles Blair was commissioned in April 1940 as a direct entry air gunner, an unusual breed, and a little bit older than the others. He's flown in the Battle of Britain. Also comparatively 'new' to the crew is a New Zealander, Pilot Officer Gilbert Redstone, flying as second pilot. He arrived from Harwell almost exactly a month ago.

David Vandervord, the officer who usually occupies the second pilot seat, is not flying that night. He's been with the crew through some tough operations, including a recent attack on a German battle cruiser in Brest and a raid on Berlin. Over Hamburg, on 13 March, he was wounded in the leg. They were hit by flak and attacked twice by enemy night fighters, the gunners both opening fire and perhaps even making a 'kill'. Their bomber (Wellington R1448) was badly damaged, so much so that one of the German night fighters claimed her as 'destroyed'. But she made it home. David is now recovering from his wounds.

The target for 25 April is Kiel. Hitting Germany's capital ships remains a key priority. They've been there before, earlier in the month. Twice. And nearly didn't make it back. On one of those raids, searchlights and fighter attacks forced them down to the deck and into the light flak. The rear turret was put out of action. It was another lucky escape.

At 20.54 hours they take off again. They are carrying a mixed load of general-purpose (GP) bombs and incendiaries. Three hours later, and the faint call of an SOS is heard and a 'fix' is taken. They are over the North Sea, about eighty miles from the Dutch coast. It is the last message they will ever send. The returning crews tell of many searchlights and intense heavy flak. There are also flak ships to catch them as they head out to sea. V-Victor is missing, never to return.

Main Image George Agar (left) and crew. David Vandervord (second left) was not flying on the night the crew went missing.

A OJ

BOMBER COMMAND AT WAR
PART ONE 1939 TO 1942

Left Crews of No. 149 Squadron walk towards their Vickers Wellington IAs at RAF Mildenhall, Suffolk, for a training sortie on 21 December 1939. The Squadron's operations at this time were mostly directed against German naval installations and shipping.

Bomber Command's offensive against the enemy began less than an hour after the declaration of war against Germany. At noon on 3 September 1939, a lone Bristol Blenheim of No. 139 Squadron took off from RAF Wyton in Huntingdonshire, to reconnoitre units of the German Fleet leaving Wilhelmshaven. During the sortie it became the first British aircraft to cross the German frontier during the Second World War. That evening, acting on information obtained by the Blenheim, a force of Hampdens and Wellingtons was despatched to attack the vessels but was defeated by bad weather. Late that night the Command made its first night foray to penetrate into the Reich, when Whitleys of Nos 51 and 58 Squadrons dropped more than six million leaflets over Hamburg, Bremen and the Ruhr, and the following day bombs were dropped for the first time in anger when Blenheims of Nos 107 and 110 Squadrons attacked German warships in the Schillig Roads, near Wilhelmshaven and Wellingtons of Nos 9 and 149 Squadrons attempted an attack against naval targets off Brunsbüttel. Within a month Whitleys were over Berlin. They released only leaflets, but in doing so sowed the seeds of a whirlwind that would manifest itself during the winter of 1943–44.

These early operations marked the culmination of preparation begun five years earlier, in July 1934, when the Prime Minister, Stanley Baldwin, announced the Government's intention of increasing the strength of the Royal Air Force (RAF) by forty-one bomber squadrons by March 1939, which it was claimed would achieve parity with the emerging threat of German rearmament. On 14 July 1936 the Home Defence Force, previously known as Air Defence of Great Britain, was reorganised into Bomber, Fighter, Coastal and Training Commands. The task facing Air Chief Marshal (ACM) Sir John Steel – the appointed Commander-in-Chief (C-in-C) of the newly formed Bomber Command at his headquarters, Hillingdon House, Uxbridge – was enormous. His force amounted to twenty squadrons, within two groups, equipped with obsolescent biplanes: No. 1 Group had Hart and Hind light bombers and No. 3 Group lumbering Heyfords and Virginias with a single day-bomber, the Overstrand. The next aircraft to enter service, the monoplane Hendon and Harrow, were little better, and it was not until the introduction of the Whitley and Blenheim with enclosed cockpits, power operated turrets, variable pitch propellers and retractable undercarriage, that Bomber Command could be considered a viable force for carrying the war to the enemy. Even so, numbers were small and despite the formation of two new groups (Nos 4 and 5) in 1937, by September of that year Bomber Command could only muster 96 bombers, against the Luftwaffe's estimated 500. These would be supplemented by the arrival of the Hampden in September 1938 and the Wellington the following month. The formation of new squadrons and the re-equipment of existing units created technical and training difficulties as personnel sought to master more complex technology. Meanwhile, the Air Ministry was developing a series of plans for the deployment of the growing force, but the Command's new C-in-C, ACM Sir Ludlow-Hewitt, was under no illusion that these would come to nothing unless the force was capable of withstanding the enemy defences and operating in all conditions. There were significant training and equipment deficiencies in relation to navigation, bomb aiming and gunnery.

The Munich crisis of September 1938 highlighted further shortcomings in organisation and logistics as Bomber Command was called to mobilise. Despite the apparent respite gained by Chamberlain, the international situation continued to deteriorate. Munich had been a wake-up call and training intensified under Ludlow-Hewitt's supervision, although marked weaknesses remained in navigation and gunnery training. Realising his force's limitations the planners now confirmed that if war was declared, targets would be restricted to military and naval objectives, unless the Luftwaffe attacked British civilian targets, although industrial targets remained an option. Plans were made for Battles and Blenheims to be based at French airfields, enabling them to attack positions in Germany if required.

Following Roosevelt's appeal to the British, French, German, Italian and Polish Governments to refrain from launching air strikes on civilians after the German invasion of Poland on 1 September 1939, British policy decreed that in the event of war, targets on German soil were inviolate, restricting attacks to ships at sea or in harbour. Flights over Germany would be limited to reconnaissance and leaflet dropping. Although humanitarian concerns prescribed this political decision, the reality remained that despite five years of growth and technical improvement, Bomber Command was still ill-prepared to mount a decisive offensive against the might of Germany.

This was soon brought home by losses suffered during the Command's initial daylight raids. On 4 September five out of fifteen Blenheims and two out of fourteen Wellingtons despatched to bomb German warships were lost to flak and fighters, becoming Bomber Command's first operational losses. In France the Command's Battles, now operating as part of the Advanced Air Striking Force (AASF) suffered even greater casualties. Attempts to reduce losses by the use of formations with the bombers' gunners providing mutual protection against fighters were soon negated as the Germans changed tactics, seeking out the bombers' blind spots. During December 1939 nineteen Wellingtons attacking naval targets at sea succumbed to enemy action, in addition to two Hampdens lost to friendly fire. Hampdens, Wellingtons and Whitleys operating by night dropped leaflets (codenamed 'Nickels') over Germany,

Whitleys being the first RAF aircraft to reach Berlin on 1/2 October. Later, flying from advanced bases in France, they ranged as far as Austria, Czechoslovakia and Poland, faring better against enemy defences. The cover of darkness confounded friend and foe alike. Such sorties provided operational experience, while highlighting the difficulties of relying on dead-reckoning navigation over a blacked-out Europe with limited radio aids.

Although certainty of position was desirable when releasing Nickels, it was a pre-requisite for accurate bombing, and essential for a safe return to base. In March 1940 the rules of engagement changed. A Luftwaffe attack on the Royal Navy at Scapa Flow caused accidental civilian casualties, prompting a retaliatory strike on 19/20 March 1940, against a seaplane base on the island of Sylt, the Command's first against a land target. The results were far from encouraging. Despite forty-one crews claiming a successful raid, most bombs missed their target and fell into the sea.

The replacement of Ludlow-Hewitt by Air Marshal Portal as Commander-in-Chief in April 1940 coincided with a move to a purpose-built headquarters at Naphill, High Wycombe, from temporary accommodation at Richings Park, Langley, taken over in August 1939. Portal immediately faced a new challenge as German forces began their advance in the west. Bomber Command was directed to mine Norwegian waters and attack shipping in an attempt to stem the invasion of Norway. Again, daylight operations suffered heavy losses and now fighters were found also to be operating against night attacks. The month also saw

Right Twin-engine Handley Page Harrow aircraft of No. 214 Squadron at RAF Feltwell in 1937. (The Banks collection)

the formation of the first Commonwealth squadron in Bomber Command, No. 75 (New Zealand) Squadron, and the first two Bomber Command Operational Training Units.

During May and June the Command focused briefly on the Norwegian campaign before concentrating on the defence of France and the Low Countries. While bombers operating with the AASF flying from French bases attacked transport and communications targets, Portal was required to allocate No. 2 Group's Blenheims and No. 4 Group's Whitleys for day and night operations to support the land battle in northern France and Belgium. At the same time Portal's desire to attack the Ruhr was being rejected by both the French and Chamberlain who feared that such a move would escalate the conflict to attacks on urban centres. Chamberlain's replacement by Churchill on 10 May and the Luftwaffe's attack on Rotterdam two days later, caused a review of policy and Bomber Command was finally granted approval to carry out bombing operations against Germany, initially targeting road and rail targets supplying the Wehrmacht's advance into the Low Countries. On 15/16 May 1940, for the first time over 100 Wellingtons, Hampdens and Whitleys were despatched. While a few were directed against Belgian communications targets, ninety-nine attacked east of the Rhine, targeting sixteen oil, steel and rail targets in the Ruhr. Immediate results appeared to justify the new policy both in terms of reduced losses and claims of success by the crews.

Raids on Italy commenced on 11/12 June, when thirty-six Whitleys operating from the Channel Islands raided Turin and Genoa. The latter objective subsequently received

Left The twin-engine Armstrong Whitworth Whitley entered RAF service in 1937 and was one of the Royal Air Force's main front line bomber types at the start of the Second World War. (The Mitchell collection.)

Below Bomber Command aircrew climbing in to their Armstrong Whitworth Whitley. (The Mitchell collection)

further attacks by Wellingtons taking off from airfields near Marseille before the RAF withdrew from French bases. However, the price of trying to stem the German advance had been high, especially among the day-bombers in France. By the end of June Bomber Command had lost half its front-line strength, including many experienced regular aircrew.

Night attacks on oil and communications targets in the Ruhr were stepped up. Crews demonstrated great determination and courage, one such raid in August resulting in the award of the Victoria Cross for Flight Lieutenant Roderick Learoyd for pressing home his attack on the Dortmund Ems Canal in the face of fierce resistance. Meanwhile, surviving Blenheim crews were tasked for cloud cover daylight raids on Germany and attacks on airfields in occupied territory.

The build-up of forces in preparation for Operation Sealion, the projected invasion of Britain, provided new targets, notably concentrations of barges in the Channel ports between July and September. It is an aspect of the Battle of Britain often overlooked, yet on 3 September Churchill recognised the bombers' contribution and perhaps more significantly their potential when he penned his minute to the War Cabinet: 'The Navy can lose us the war, but only the Air Force can win it. ... The Fighters are our salvation ... but the Bombers alone provide the means of victory.'

If this was so, it was going to be a long hard road for Bomber Command. An accidental Luftwaffe attack on London on 24/25 August prompted a retaliatory strike against industrial targets in Berlin the following night. While potentially headline making, it proved to be little more than a gesture – with Wellingtons, Whitleys and Hampdens attacking independently, releasing minimum bomb loads and then struggling to regain base in the face of deteriorating conditions. Of the eighty-one aircraft despatched, twenty-nine claimed to have bombed the target and five failed to return. A month later Portal was ordered to cause 'Germany's economic disruption' but the intention was still to hit specific targets within urban areas, rather than conduct indiscriminate bombing. Portal was sceptical of his Command's ability to achieve such a task, but lacking photographic evidence he was unable to convince the Air Staff that their intentions were unrealistic. However, soon he was promoted to Chief of the Air Staff, and on 5 October 1940 the problem passed to his successor, Air Marshal Sir Richard Peirse.

Under Pierse the Command began to obtain the photographic evidence of the results of attacks, which indeed confirmed not only the difficulty of finding a specific target within an urban area, but also on occasion of even identifying conurbations accurately. By continuing Portal's policy Peirse was by default initiating area attacks. On 16/17 November 1940, when he despatched130 aircraft – the largest force to date – against Hamburg, only 60 claimed to have bombed the target. When tasked with concentrating on oil targets in January 1941 Peirse instigated a policy of attacking these during the better visibility of the full moon – the so-called 'moon period' – while making a major attack on industrial cities during the new moon. By late February his instructions had changed again: the Battle of the Atlantic was raging. Capital ships, U-boats and the long-range Focke-Wulf Condor patrol aircraft, their bases and infrastructure became his objectives. These not only encompassed coastal positions, which were easier to find and identify, but provided opportunities for attacks against large ports, such as Hamburg and Kiel, in addition to the oil targets.

Wellingtons were now beginning to re-equip some of the former Blenheim squadrons and becoming the backbone of the Command. The first of the four-engined heavy bombers, the result of specifications drafted by the Air Ministry in 1936, commenced operations: the Stirling in February 1941 with the Halifax the following month. Invariably they experienced teething problems, as did the Manchester, Avro's response to a specification of the same period, and production was slow. At the same time heavier and more destructive bombs were being introduced, the first 4,000lb Cookie being dropped by Wellingtons on Emden on 31 March 1941, and the first sign of further expansion of the bomber force came with the formation of Royal Canadian Air Force (RCAF) units making their operational debut in June, to be joined shortly by those from the Royal Australian Air Force (RAAF).

Meanwhile, the Blenheim squadrons were tasked with anti-shipping sweeps off the French and Dutch coasts, together with low-level attacks on specific military and industrial targets. Such targets were becoming increasingly well defended by flak and casualties were heavy. An attack by fifteen Blenheims against Bremen on

Left The twin-engine Bristol Blenheim light bomber which was operational with Bomber Command squadrons during the early years of the war, notably during the Battle of France and attacking enemy barge concentrations in the Channel ports during the Battle of Britain.
(The Christian collection)

Opposite Bristol Blenhiem R-GB T1989 of No. 105 Squadron.
(The Christian collection)

4 July resulted in the award of the Victoria Cross for its leader, Wing Commander Hugh Edwards. The following month fifty-four Blenheims, supported part-way by large numbers of aircraft from Fighter Command, executed the deepest daylight low-level penetration raid to date, attacking two power stations near Cologne. Throughout this period the light bombers, and on occasion some heavies, were despatched with fighter escort on so-called shallow penetration 'Circus' raids in an attempt to engage and deplete the Luftwaffe in combat. It was an ill-fated strategy and enemy fighters continued to remain a tangible threat to daylight operations over France without the benefit of cloud cover.

In July 1941, Peirse was formally instructed to 'direct the main effort of the bomber force ... towards dislocating the German transportation system and to destroying the morale of the civil population as a whole and of the industrial workers in particular'. For the next four months attacks were conducted against rail yards surrounding the Ruhr on moonlight nights, while on dark nights the bombers concentrated on cities along the Rhine, or when weather prevented these, on other German city targets. The increased effort and greater tonnage of bombs now being dropped on Germany was offset by increased casualties as the German defences were strengthened and reorganised, particularly in respect of the night-fighter force.

On 7/8 November 1941, 169 aircraft were despatched to Berlin; only 73 claimed to have reached the target, causing fires on the outskirts of the city. Some 21 aircraft failed to return. An additional 7 Wellingtons from a force attacking Mannheim were lost, together with a further 9 aircraft on other missions, making this Bomber Command's most costly night to date. While poor weather was attributed as a contributory factor, doubt was also cast over Peirse's management of the bomber offensive and the War Cabinet decreed that bomber operations should be severely restricted over the winter months pending a review of policy.

However, the bomber offensive was already under scrutiny. In August 1941 the Butt Report, a survey of over 600 night bombing photographs, commissioned by Churchill's chief scientific adviser, had concluded that of those aircraft recorded as attacking German targets in general, only one in four got within five miles. Over the Ruhr it was only one in ten and on dark nights it was only one in fifteen. Taken overall, only about one-third of aircraft claiming to reach their target in reality achieved this and only five per cent actually bombed within five miles. The report was challenged by the Air Staff who, in September, presented their own proposal for improving the efficacy of the bomber offensive. Based on an analysis of damage to UK cities by the Luftwaffe, the proposal calculated that given a force of 4,000 bombers and using new navigation devices under development, the RAF would be able to destroy 43 German towns with a population exceeding 100,000. By doing so, Portal argued, it would take six months to win the war. Churchill was sceptical of this claim, and rejected the request for 4,000 bombers, but agreed to maintain his support for Bomber Command at its present strength and the continuation of present policy. Nevertheless, for more than three months the bomber offensive was severely restricted, with attacks limited to French and German ports, notably Brest, Lorient, Hamburg, Emden and Bremen, plus occasional targets in the Ruhr. No. 2 Group continued to operate their Blenheims in low-level daylight attacks, provided support for the first Combined Operations raid against Vaagso on 27 December, and that night initiated the first Intruder raids against enemy airfields in a renewed effort to deplete Luftwaffe activity.

The year 1942 witnessed the entry of the United States into the war against Germany, bringing new military might and industrial resource into the European theatre. Early on it also marked a sea change for Bomber Command in respect of leadership, strategy and equipment. Air Marshal Peirse was replaced, temporarily, in January by Air Marshal Jack Baldwin. Baldwin was given a directive to focus operations on the morale of the enemy civil population and in particular of industrial workers, thereby formalising the activity previously implemented by Portal and Peirse.

Deliveries of the third of the four-engined heavy bombers, the Avro Lancaster, had begun on Christmas Eve 1941. Arising like a phoenix from the ashes of the unreliable Manchester, this would soon establish itself as the Command's foremost aircraft. A new radio navigation aid, Gee, was introduced, which would help crews concentrate their attacks against industrial targets at least as far as the Ruhr, but time was of the essence. It was anticipated that Gee would only have a useful life of about six months before the enemy developed countermeasures to reduce its effectiveness.

Baldwin's new strategy focused on the German industrial heartland. Essen, Duisburg, Düsseldorf and Cologne were selected as primary targets, while the north German ports of Bremen, Wilhelmshaven and Emden were secondary ones. Further objectives, beyond Gee range, including Berlin, would be attacked in suitable weather conditions. Since Gee was only a navigation and not a bombing aid, targets were now defined as built-up areas rather than specific factories or dockyards. Determined out of expediency, the strategy was driven by operational practicalities rather than strategic or moral considerations.

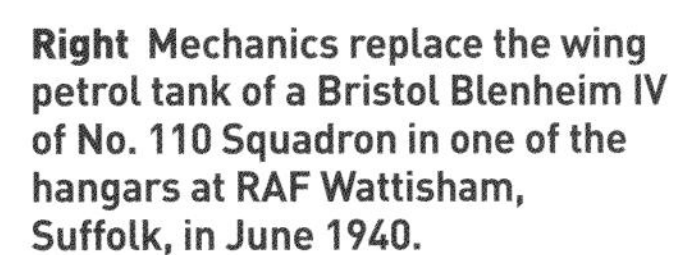

Right Mechanics replace the wing petrol tank of a Bristol Blenheim IV of No. 110 Squadron in one of the hangars at RAF Wattisham, Suffolk, in June 1940.

Far right The twin-engine Handley Page Hampden medium bomber, often referred to by aircrew as the 'Flying Suitcase', was flown operationally with Bomber Command during the early years of the war. The aircraft in the picture shows the VN squadron code indicating it is in service with No. 50 Squadron. (The Dawson collection)

Nevertheless, despite the continued development of technical aids that would refine bombing accuracy, this strategy would largely dictate the conduct of the Command's operations for the remainder of the war.

However, further questions were already being asked in political circles about the validity of the bombing campaign and how much of the country's manufacturing resources should be allocated to the production of aircraft and equipment for the Command in the face of competing demands. Strong rebuttal was required if the bomber offensive was not to be diluted. It was therefore crucial that Baldwin's successor, Air Marshal Arthur Harris, appointed in February, should be a forthright and resolute champion of the efficacy of area bombing and its potential to undermine the ability of the enemy to continue fighting. Lord Cherwell, who had commissioned the Butt Report, was also a strong advocate of the new strategy. He provided a mathematical projection, which predicted that half the bomb loads of a force of 10,000 bombers directed against 58 major German towns would render one-third of their population homeless. Despite being an optimistic and impractical thesis it was influential in maintaining support for Bomber Command and its new commander.

In reality Harris's force of 446 medium and heavy bombers suitable for night attacks, most of them Wellingtons, was too small to achieve such an effect. Cherwell's predicted force was beyond the capability of both the production of aircraft and the training of crews, and the enforced transfer of men and equipment, including whole squadrons, to support the Middle East offensive and Coastal Command, continued to hobble Bomber Command's expansion. However, the continued conversion of No. 4 Group's Wellington and Whitley units to the Halifax, and the replacement of No. 5 Group's Hampdens with the Lancaster, significantly increased the weight of bomb load that could be delivered per aircraft.

Harris believed that the efficient and effective use of his force required concentrated attacks directed against cities, rather than the spasmodic raids by smaller forces as conducted during 1941. Faced with only slow growth in the size of his Command he was determined to increase the effectiveness of attacks. In less than a fortnight from his appointment, on 3/4 March 1942 he sent some 235 aircraft to strike the Renault works at Billancourt on the outskirts of Paris. Attacking from a lower level than usual in the light of flares dropped by the raid leaders navigating by Gee, the aircraft achieved the highest level of concentration yet, causing significant damage for the loss of only one aircraft to enemy action. A less successful attack on Essen five nights later by 201 aircraft again used raid leaders navigating for the first time with Gee. But dropping flares at which the following force aimed their high-explosives, revealed that although a vital development, Gee was not the solution to all the Command's problems.

The success of low-level precision attacks against specific targets prompted Harris uncharacteristically to attempt similar, longer-range daylight operations using his heavy force. Tasked to support the anti-submarine war

Opposite Photograph taken from a No. 4 Group Halifax, showing a daylight raid by 47 aircraft on Brest docks in progress on 18 December 1941. Hits on the German warship 'Gneisenau' were reported but six aircraft failed to return. The photograph is from the Cavalier collection, sourced from the photographic section at RAF Middleton-St-George.

Above right Taken from the Clydesmith collection this picture of Vickers Wellington HA-A R1008 had the caption, 'Targets flown Mar 41 to July 41, Brest, Hamburg, Berlin, Essen, Düsseldorf, Kiel, Cologne, Bremen.'

Right Pilot David Donaldson, third from the left and with his crew in front of their Vickers Wellington, had a distinguished career serving with Bomber Command. (The Grundy collection)

he was already despatching his force to lay mines from the Bay of Biscay to the Baltic Sea. On 17 April 1942 the MAN submarine engine factory at Augsburg was targeted by Lancasters of Nos 44 and 97 Squadrons, and on 17 October the Schneider armament works at Le Creusot by a larger force of ninety-seven Lancasters of No. 5 Group. Such operations again proved that long-range daylight operations by small numbers of heavy bombers were both impractical and costly; of the fourteen aircraft sent to Augsburg only five returned. 'Moling' operations – despatching single aircraft to attack targets in daylight using cloud cover for concealment – proved of limited success, but at best could only be of nuisance value and often resulted in early returns when the cloud cover disappeared before the target was reached.

While Gee addressed the problem of navigating to the target, it did not resolve the issues relating to the identification and marking of an aiming point to ensure the accurate delivery of bomb loads. Bomb aimers still relied on the visual identification of their aiming point, which might soon become obscured by smoke even in fair weather, with the result that many bomb loads were missing their intended target.

The use of incendiary bombs to illuminate the aiming point for following crews was employed with encouraging success by 234 aircraft against Lübeck on 28/29 March. Outside Gee range, the objective's coastal position simplified navigation and the port's light defences permitted bombing from a lower level. With a large percentage of closely packed timber-framed buildings, Lübeck proved an ideal incendiary target. The Ruhr, however, was a totally different proposition. Harris was determined to demonstrate that similar tactics could be applied even in the face of heavy flak and searchlight glare. He persuaded Churchill and Portal to commit not only all his front-line force but also aircraft and crews from training units to mount a 1,047-aircraft attack against Cologne on the night of 30/31 March. The attack was completed in ninety minutes by concentrating the aircraft in a stream that saturated the defences and minimised losses. Gee-equipped leader aircraft dropped flares to guide the following crews, who in turn dropped incendiaries to create fires at which the final crews released their high-explosives. The raid was a resolute success, but at a cost of forty-one aircraft.

Attempts to repeat the scale and tactics of attack against Essen and Bremen in the following weeks met with mixed success. Nevertheless, Harris had made his case and raised the standing of Bomber Command. Despite the losses, these attacks demonstrated the potential of a large concentrated bomber force and the importance of tactics, but for the time being such a large force was unsustainable.

There was also the issue of making it to a target in the face of fierce opposition. The new tactics had demonstrated experienced crews using Gee could find and bomb the objective accurately. Group Captain Sydney Bufton, Director of Bomber Operations at the Air Ministry, favoured the formation of a specialist unit of experienced crews who could be trained in the latest techniques and given the best and most up-to-date equipment. Harris, who was against the formation of an elite unit, was in favour of competitive selection using the finest crews from each squadron as raid leaders for each of his groups. After fierce debate Bufton, supported by Portal, won the day and in August

Far left On her wedding day, and surrounded by Polish aircrew, is Wanda Szuwalska, third from the left. (The Szuwalska collection)

Left Vickers Wellington LF-D N2992 with a 'dustbin' turret protruding from the belly of the aircraft. (The Banks collection)

Harris created his 'Pathfinder Force', under the command of an outstanding Australian airman, Air Vice Marshal (AVM) Donald Bennett.

Initial Pathfinder attacks were disappointing until a new technique was adopted, using a vanguard of 'illuminators' dropping white flares, followed by 'visual markers' dropping coloured flares to indicate the proximity of the target and acting as an aiming point for subsequent 'backers-up' with incendiary loads. By September the first specialist marker bomb – known as a 'Pink Pansy' – was being employed. Flares and coloured markers assisted in concentrating the force over the target area, but did little to improve bombing accuracy. Again technology provided the answer. Introduced in December, Oboe was a radar device that permitted single aircraft to release their bombs with unprecedented accuracy. In short supply and only able to handle one aircraft at a time, Oboe was fitted to Mosquitoes and along with the development of new target marker bombs, gave the following bomb aimers a clearly defined aiming point for visual bombing.

While new technology was providing solutions for Bomber Command, the Germans were developing equipment and techniques to make the skies over Germany increasingly more dangerous for the bomber crews. Increasing numbers of radar-equipped night fighters were being vectored on to the bomber stream by controllers monitoring a series of radar-controlled 'boxes'. Radar-directed flak and searchlight belts surrounded major targets, notably those in the Ruhr. Numerically too small to saturate these enhanced defences Bomber Command's losses began to rise again, particularly among the Halifax and Stirling squadrons.

For daylight operations the Blenheims of No. 2 Group were now being replaced by American types, the Boston and Ventura, with the remarkable twin-engined Mosquito equipping two squadrons and further deliveries to follow. This Group specialised in the 'Circus' raids, together with intruder sorties against enemy airfields and precision attacks such as Operation Oyster, in December 1942, targeting the Philips factory at Eindhoven that was producing vital electronic components, including valves essential for German radar.

By the end of 1942, however, Bomber Command was at the end of a year that had seen the genesis of new tactics for the employment of the heavy night bomber force. Aided by new navigation, blind bombing and target-marking equipment, they were now poised to launch the type of attacks originally proposed two years earlier. The US 8th Air Force had arrived in England during the summer of 1942 to establish itself on airfields in East Anglia. Despite British scepticism it had begun to fly initial daylight operations against targets in occupied Europe. As confidence grew it prepared to extend them deep into Germany. The bomber offensive was growing in strength. Already the British and American Chiefs of Staff were committed to a combined bombing assault that would debilitate the German war machine and create the conditions necessary to permit a successful land invasion of the Continent.

LIFE AND DEATH IN A FLYING SUITCASE

The night of 11 May 1940 is a significant date in the history of No. 49 Squadron, and of one of its young pilots, Pilot Officer David Drakes. Six of the squadron's Handley Page Hampden bombers are detailed for operations to München-Gladbach, their first attack of the war on a German target. Winston Churchill has been Prime Minister for a day and the 'Phoney War' has fast given way to a new reality as German forces sweep through Holland. Bomber Command has a chance to prove its worth by bombing road and rail communications supporting the German advance.

The attack appears to go well, most believing they have hit the target. Each of the Hampdens is carrying four, 1,000lb instantaneous bombs, and the crews are confident of success. But not all is as it should be. Flak is heavy, dirty puffs hiding deadly concentrations of danger, spitting out red hot shrapnel into the night sky. An Armstrong Whitworth Whitley V of No. 77 Squadron is hit and crashes to earth, the first to be lost in Germany while carrying out a bombing operation on a mainland target. A Hampden is also hit on its bombing run; the skipper, Wing Commander Arthur Luxmoore, orders his crew to bail out. Three take to their 'chutes before the 'flying suitcase' comes down and the wing commander is killed in the crash. They had been flying their eighth operation together.

On the journey home, Drakes is in trouble too. The starboard engine splutters and dies as they reach the French border, and the aircraft steadily loses height. With no hope of making it across the Channel, Drakes skilfully brings the aeroplane down on the coast, near Le Tréport. He and his crew emerge unscathed to return to Scampton. Six months later, and now promoted to flying officer, Drakes flies to Berlin for his thirty-ninth operation, completing his first tour.

In July 1941, Drakes returns to No. 49 Squadron to begin his second tour. It is largely a new team, although he has persuaded the wireless operator/air gunner (WAG) from his first crew, Bill Watson, to join him. War accelerates promotion: Bill started operations as a leading aircraftman (LAC) and now he is a flight sergeant; Drakes is a squadron leader, and still only twenty-two years of age.

On the night of 1 November, Drakes is briefed for his forty-eighth operation. He is with his co-pilot, Pilot Officer Victor Beaney, and Observer, Pilot Officer William Cheetham. They are to be in the company of three other Hampdens, detailed for minelaying and anti-shipping operations. Drakes is tasked with the latter, and takes off shortly after 22.00 hours to seek suitable targets off the Frisians.

By the small hours of the following morning it is clear there has been trouble. One Hampden has already returned with a wounded rear gunner. As dawn breaks, there is no sign of Drakes and his crew. They have disappeared, nothing being heard or seen of their aircraft after take-off. Hampden AE224 has failed to return, and the words 'missing' chalked on to the operations blackboard. They are one of hundreds of aircrew to disappear without trace, the victim of flak or a night fighter over the cold North Sea. Their bodies are never recovered.

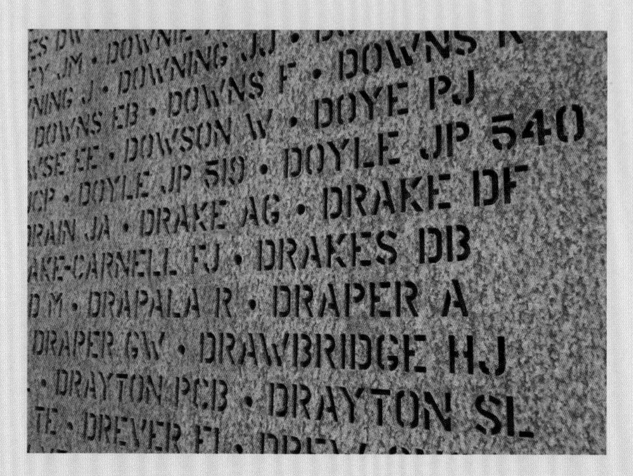

Left The loyal groundcrew on the wing of a No. 49 Squadron Hampden.

Right David Drakes (second from left) survived an early scare to be lost on his 48th operation.

ON

BOMBER COMMAND AT WAR
PART TWO 1943 TO 1945

Left A four-engine Avro Lancaster of No. 61 Squadron about to be loaded with a general bomb load. The Lancaster became one of the most successful and versatile aircraft to serve with Bomber Command during the Second World War. The picture is from Don Watson's collection, taken when he was serving with the squadron.

By the start of 1943 Bomber Command was poised to intensify the offensive. The Pathfinder Force, now redesignated No. 8 (Path Finder Force) Group (PFF), was now equipped with new bespoke Target Indicator bombs and No. 6 (RCAF) Group, formed in October 1942 from Canadian squadrons in No. 4 Group, began full-scale operations.

In January the Allied Combined Chiefs of Staff debated the bomber offensive at the Casablanca Conference. A revised directive was issued. Bomber Command was to conduct area attacks by night while the Americans made precision attacks by day. Their joint objective was now 'the progressive destruction and dislocation of the German military, industrial and economic system, and the undermining of the morale of the German people to the point where their capacity for armed resistance is fatally weakened'. However, just as it seemed that Harris would be able to increase his onslaught against Germany, the Battle of the Atlantic intensified. With U-boats causing increasing shipping losses, Harris was ordered to switch his emphasis to their Biscay bases. The attacks were of only limited value, since any vessels in port were berthed in substantial shelters. Nevertheless, operations were also conducted against German cities including Berlin, Cologne and Hamburg, with long-range sorties over the Alps to continue attacks on north Italian targets begun the previous autumn.

A further navigation device, H2S, was introduced in small quantities in January. An aircraft-mounted, ground-scanning radar, this was independent of ground stations and did not suffer the range limitations of Gee. Since H2S produced a radar map of the ground beneath the aircraft it could 'see' through cloud and had the potential to be used as a blind bombing aid in addition to its primary navigation role. As with Gee and Oboe, however, it was not a cure for all Bomber Command's problems, being best suited to targets with high-contrasting radar reflectivity, such as those near water or woods. Wilhelmshaven was attacked to great effect on 11/12 February, while specific areas of large homogenous urban centres such as Berlin proved harder to identify with any certainty.

By March Bomber Command comprised some sixty-five squadrons, of which three were Special Duties squadrons not directly involved in the main bombing offensive. Harris was able to consider switching his emphasis from the French ports back to Germany. The Ruhr, Germany's industrial heartland, became his main target for sixty per cent of attacks, enabling the use of Oboe for blind marking since there were still few H2S sets available. Other operations were against targets dispersed as far as Aachen and Pilsen, Stettin and Turin, thereby discouraging the enemy from concentrating all his defences in the industrial belt, while new crews gained experience with minelaying sorties. The Battle of the Ruhr, as it became known, subjected key manufacturing towns to twenty-four major raids in four months – a total of 18,506 sorties, from which 827 aircraft failed to return.

May saw the audacious attack on the Möhne, Eder and Sorpe Dams by nineteen specially modified Lancasters of No. 617 Squadron, each carrying Barnes Wallis's 'bouncing bomb' known as 'Upkeep'. Operation Chastise was intended primarily to destroy supplies of water essential for industrial processes. The breaching of the dams also served to deprive the working population of its domestic supply and caused additional collateral damage to buildings and communications infrastructure. The water ruined electrical installations and machinery – each effect contributing to the strategic aim of reducing industrial output and undermining morale. The value and effectiveness of the Dams Raid has been an issue of much subsequent debate, yet the vital significance of the dams to the German war machine is irrefutably confirmed by the high priority and extreme efforts made by the Germans to repair them by the autumn of 1943 to ensure sufficient water for production in 1944. Regardless of the success of the operation it came at a high price. Eight aircraft failed to return, all but three of their crew members being killed.

Once again Harris's aversion to the creation of a specialist unit, and 'panacea targets' had been overridden by a direct instruction by Portal. However, Harris soon appreciated the merits of such a formation. The new squadron had increased the size and efficacy of his force. It comprised experienced aircrew who might otherwise have been instructing, rather than carrying the war to the enemy, and also prevented having to remove a Main Force squadron from the front line for a protracted period of training. Moreover, the operation had not only demonstrated Bomber Command's ability to conduct successful precision strikes against small and well-defended targets deep in the German heartland by night, but also pioneered the use of VHF radio to direct the bombing force during an attack, thereby permitting tactical flexibility to improve the effectiveness of the raid.

During the opening months of 1943 No. 2 Group commenced operating Mitchells, while continuing to despatch Bostons, Venturas and Mosquitoes against precision targets. A new tactic was introduced: nuisance raids conducted by individual Mosquitoes against objectives such as Bremen, Hamburg and Wilhelmshaven, disturbing the sleep of their populations. On 30 January, Mosquitoes flew in daylight to Berlin, interrupting speeches by Goebbels and Goering, and in May Mosquitoes mounted a deep-penetration daylight raid on the Schott optical works at Jena. The following month No. 2 Group, with the exception of Mosquito squadrons Nos 105 and 139, was transferred to the Second Tactical Air Force. In addition, three Wellington squadrons were sent to the Middle East

Opposite Four-engine Handley Page Halifax DT-A of No. 192 Squadron, part of No. 100 Group which specialised in electronic warfare and countermeasures. (The Grundy collection)

Opposite below A four-engine Short Stirling of No. 90 Squadron. The Short Stirling was the first of the four-engine heavy bombers to enter RAF service. (The Collyer-Smith collection)

for five months before returning once more to the UK and the night offensive against Germany.

A modification to the Casablanca Directive in June saw the emphasis switch to factory, infrastructure and airfield targets, intended to degrade the Luftwaffe's fighter force. Increased use of H2S was now being used by the PFF. This and the introduction of the Master Bomber technique pioneered on the Dams Raid, would increase the accuracy of marking and enable the point of aim to be changed during an attack, thereby ensuring more effective distribution of bomb loads.

Harris now changed his attention from the Ruhr to other German cities beyond Oboe range, notably Hamburg. This port and industrial centre was subjected to intense attack from Bomber Command by night and the United States Army Air Forces (USAAF) by day. The offensive opened on the first night of 24/25 July with 728 aircraft using H2S and visual bombing, causing severe damage. During this attack the bombers utilised to great effect a new weapon devised to reduce losses – 'Window' – being strips of metal foil that produced a myriad of false echoes to confuse the enemy's radar-controlled night fighters and defences. American attacks during the following two days were less successful; smoke from fires started the previous night prevented bomb aimers from seeing their precision targets. Over 700 aircraft of Bomber Command flew back to Hamburg on the night of the 27th to deliver a well-concentrated attack. Numerous fires were started in residential districts. These rapidly merged into a raging inferno, sucking in oxygen from surrounding areas to create a firestorm lasting nearly three hours and resulting in the destruction of some 16,000 buildings and the loss of approximately 40,000 lives. Bomber Command returned again on the nights of 29/30 July and 2/3 August. As if this were not sufficient, small numbers of Mosquitoes carried out nuisance raids on the city during the intervening nights.

The end of July and beginning of August saw 'shuttle raids' against targets in northern Italy as a precursor to the Allied invasion, with aircraft landing at North African bases to refuel and re-arm before bombing other Italian targets on their return route to the UK. The Master Bomber technique, pioneered on the Dams Raid was used for the first time by nearly 600 aircraft of Main Force in an atypical precision attack against the German research establishment at Peenemünde on 17/18 August during the period of the full moon. Although successful, a high price was paid. The Germans had quickly recovered from the setback of 'Window', instigating a system of freelance single-engined fighters – 'Wild Sau' – operating independently without radar guidance, and for whom moonlight conditions were ideal.

Encouraged by the attacks against Hamburg, as autumn approached Harris turned his attention towards Berlin. However, being beyond Oboe range, and a poor H2S target, three initial attacks at the end of August and beginning of September were disappointing. Casualties were again beginning to rise with the introduction of 'Wild Sau' and the use of upward-firing armament, known as 'Schräge Musik'. The latter allowed the enemy to take up position behind and below to fire into the bomber's vulnerable fuel tanks and bomb bay, a tactic concealed

Left A twin-engine Douglas Boston of No. 107 Squadron, which operated as part of Bomber Command's No. 2 Group. The picture, is captioned in the Cavalier collection as having been taken in September 1942.

Opposite A propaganda leaflet dropped over Italy proclaiming, 'The Bombs and the Truth'. (The Rennison collection)

LE

BOMBE

E LA

VERITA'

LE necessità della guerra ci obbligano a colpire le risorse del nemico ovunque si trovino; altrimenti queste risorse saranno usate contro di noi.

Noi dobbiamo colpire i centri ove il nemico organizza le sue attività belliche ed industriali contro di noi, gli uffici militari e governativi, i magazzini di merci che sostengono lo sforzo bellico. Se le vostre officine contribuiscono alla produzione bellica noi dobbiamo colpirle: se le vostre ferrovie trasportano truppe, armi e munizioni noi dobbiamo colpirle.

Finchè l'Italia continuerà ad essere nel campo nemico noi saremo costretti a bombardare gli obbiettivi industriali e militari italiani. Non siamo forse costretti a colpire, con la cooperazione di aviatori francesi, belgi, olandesi e norvegesi gli obbiettivi nemici in Francia, in Belgio, in Olanda, in Norvegia?

Noi non abbiamo il sadico desiderio tedesco di far strage della popolazione civile, e miriamo ad obbiettivi di guerra; ma inevitabilmente le zone contigue sono in pericolo e possono essere colpite.

Confrontate le cifre delle perdite britanniche tra la popolazione civile con quelle sofferte dalle vostre città. Tenete presente che spesso le nostre incursioni sono più violente per peso e numero di bombe di quelle a suo tempo eseguite dalla Luftwaffe sulla Gran Bretagna. Considerate tutto ciò, e traetene le vostre conclusioni.

by the use of non-tracer ammunition. As a result, targeting was switched to a more diverse selection of cities including Mannheim, Hannover, Kassel, Frankfurt, Stuttgart, Munich and Leipzig, with occasional visits back to objectives in the Ruhr. 'Spoof' raids by small numbers of aircraft against secondary targets were now introduced in an attempt to split the German fighter defences. Although not all were successful, these operations provided increased experience for crews and enabled refinement of the new marking and control techniques.

At the same time, Harris's force began a metamorphosis. In October 1943 the Wellington, which had been the backbone of the offensive since the beginning of the war, was withdrawn from Main Force operations, as was gradually the Stirling, the first of the four-engined heavies, and early marks of its stablemate, the Halifax. These aircraft were relegated to minelaying and training and replaced by increased numbers of Lancaster bombers with smaller quantities of a new and improved mark of Halifax. Harris lobbied hard for an all-Lancaster force, but the practicalities of production meant that this was never a realistic option. The factories were already fully stretched in manufacturing sufficient numbers of both types in order to replace losses and older marks, without incurring further deficit by switching firms from Halifax to Lancaster production. Gee-H, a blind bombing system, was introduced, firstly for Mosquitoes, but later seeing increased use by Nos 3 and 6 Groups.

By mid-November Harris was ready to return to Berlin. For the next four and a half months half of Bomber Command's attacks would be directed against the German capital, the remainder being deep-penetration flights to more distant cities, taking advantage of the longer nights. The following period saw significant evolution of tactics as each side sought a means to counter the other in a deadly game of cat and mouse. While 'Window' might conceal individual aircraft in the stream, it served to highlight the stream itself, enabling twin-engined night fighters already airborne and circling radio beacons to be vectored on to it as it approached and departed the target – a technique known as 'Tame Sau'. They were assisted by other aircraft dropping flares to indicate the stream's position and track while ground controllers provided a running commentary on the bombers' movements. Once contact had been made the result depended on the skill of individual crews. Indirect routing was used to confuse the German defences as to the bombers' final destination. By April 1944 Main Force would be split and two or three targets attacked during the same night, thus dividing the fighter response. The length of the bomber stream was reduced, as was time spent over the target. Smaller forces, often employing crews under training, were despatched across the North Sea as a spoof force, turning back before crossing the enemy coast, while intruder sorties were mounted against Luftwaffe fighter airfields. Radio countermeasures were increasingly employed to confound the defences – leading to the formation of No. 100 Group specifically for this task. 'Mandrel' – a radar jamming system – delayed detection of the approaching force from enemy radar. Mosquitoes fitted with 'Serrate' to detect German night-fighter radar

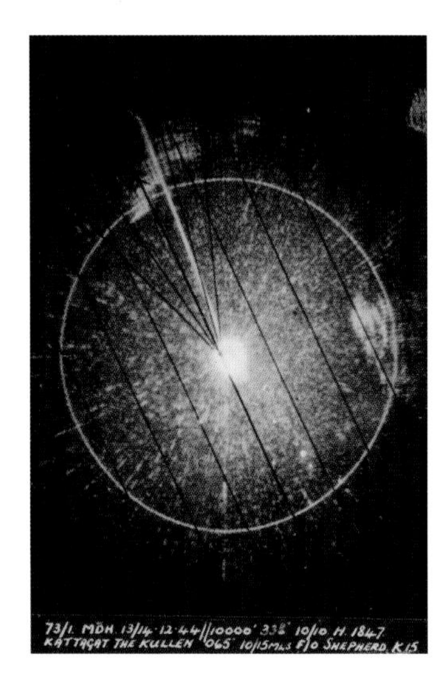

transmissions operated in the proximity of the bomber stream so that the hunter might become the hunted.

Flak and fighters were not the only foe with which the crews had to contend. The weather also took its toll: on 16/17 December 1943 the German defences accounted for twenty-five aircraft, while over England twenty-nine planes returning from Berlin crashed or were abandoned, unable to land at fog-covered bases. On 24/25 March 1944 excessively strong winds encountered during the last attack of the Berlin offensive contributed to the loss of seventy-two aircraft, eight per cent of the force.

Harris had told Churchill that the offensive against Berlin might cost his Command 400–500 aircraft, but it would lose the Germans the war. By March 1944, sixteen raids had seen Bomber Command sacrifice 500 aircraft, yet despite receiving the same tonnage as Hamburg and incurring 7,000 casualties, the population of Berlin showed no sign of capitulation. While the aircrews' courage and determination cannot be questioned, the Battle of Berlin can only be determined a costly failure.

The winter of 1943–44 made other demands. Attacks on the first V-1 flying bomb launching sites identified in northern France commenced in December. These targets were small and required precision strikes. Anxious not to dilute the offensive against Germany, Harris allocated these to Stirlings and the Lancasters of No. 617 Squadron who were not operating with Main Force. The Oboe marking was insufficiently accurate and the raids achieved little success, these compact targets subsequently being allotted to the USAAF and 2nd Tactical Air Force to attack by day.

Sorties continued during February and March against other German cities including Schweinfurt (complementing the USAAF raids on the ball-bearing industry), Augsburg, Leipzig and Stuttgart. Despite new tactics the German defences were becoming increasingly effective, scoring their greatest success on the night of 30/31 March 1944 when 95 aircraft out of 795 despatched to Nuremberg failed to return. Once again inaccurate weather forecasting was a major contributor, combined with an astute German controller who ignored smaller diversionary attacks and positioned his fighters astride the bombers' route.

It was perhaps fortunate that preparations for the proposed invasion – Operation Overlord – demanded a switch in emphasis at this point from heavily defended German cities to lightly defended targets in occupied France. Harris protested against this change of emphasis, maintaining that his crews lacked the accuracy required to prevent high casualties among French civilians, but he had little option but to comply. In the event Harris's fears were largely unfounded. French civilian casualties were inevitable, however they never reached the scale feared. After attacks on the V-weapon sites revealed the shortcomings of Oboe, Wing Commander Leonard Cheshire had pioneered a method of low-level marking. Used initially by No. 617 Squadron against French factory targets, the technique was adopted by No. 5 Group who, operating in a semi-independent role, created their own marker force. No. 3 Group evolved a technique using Gee-H, while No. 8 Group continued to use Oboe and H2S.

Although General Eisenhower took control of the RAF and USAAF strategic bomber forces on 14 April, Bomber Command continued to attack German industrial targets until the end of the month, with effective strikes on Cologne, Düsseldorf, Munich and Friedrichshafen. Attacks against French objectives, initially marshalling yards and ammunition dumps, commenced at the beginning of April. Against these lightly defended targets, involving less time over enemy territory, the average loss rate reduced. An exception occurred on the night of 3/4 May 1944, when communications problems during a raid on a tank depot at Mailly-le-Camp resulted in Main Force orbiting in bright moonlight awaiting the order to bomb. Enemy fighters arrived and 42 of the 346 attacking Lancasters were shot down. May also saw the campaign extended to include airfields and coastal batteries, with heavy attacks on myriad targets in addition to those near the intended 'Overlord' beaches to avoid any indication of where the planned invasion would land. On 5/6 June, the night of the invasion, over 1,000 aircraft of Bomber Command attacked ten coastal gun batteries while others contributed to spoof and deception operations designed to delay identification of the real assault. With the troops ashore, attention reverted to transportation targets to prevent the movement of German reinforcements to the beach-head area.

Bomber Command was a strategic force, not best suited to provide tactical support for ground forces. Eisenhower was keen to strike at German oil reserves and production facilities, allocating Bomber Command defended targets in the Ruhr. Policy was based on expediency. Between

Far left From the Moore collection and annotated, 'Mine release photo on H2S taken by me 14/12/44. Arrows are on Kullen Point.'

Left Reconnaissance photograph showing the breach in the Eder Dam, following the Dam Busters raid of 16/17 May 1943. (The Cruikshank collection)

June and August the tempo of operations increased as Bomber Command addressed immediate tactical requirements: the continued disruption of transportation and raids on fortified positions ahead of the advancing troops; the removal of the E-boat threat to cross-Channel convoys by attacks on Le Havre and Boulogne; and countering the V-1 offensive that had begun on 13 June by attacking launch sites and storage depots. Daylight bombing with fighter escort was now possible against these targets, improving accuracy and increasing the effectiveness of the attacks. A new technique emerged for Gee-H and Oboe operations whereby aircraft fitted with the equipment led small formations of aircraft that bombed when they saw the lead aircraft release its load. Nevertheless, Harris still considered his force best employed in strategic area attacks rather than against tactical precision targets. Although night raids against German oil production were conducted during this period, the emphasis remained on support for the ground forces. For the time being it would be left to the Mosquitoes of the Light Night Striking Force, formed to conduct nuisance raids in the winter of 1943–44, to continue the offensive against German cities.

By August the advancing armies were overrunning the French V-weapon sites, releasing the Allied bomber forces to switch their efforts to airfields and oil installations. In an effort to improve the effectiveness of these attacks, Bomber Command recommenced daylight operations over the Reich for the first time in three years. The following month Eisenhower's control of the bomber

Left From the Falgate collection, a study of a Bomber Command rear gunner in his turret. In training manuals the air gunners were called upon to defend their aircraft by being both the eyes and the sting.

Right Bomber Command navigator Len Dorricott, standing back left, and his No. 576 Squadron crew. Len took part in Operation Manna, the dropping of food supplies to the starving Dutch population, 'It was a marvellous feeling, the best thing I did in the war.' (The Dorricott collection)

Below right A destroyed V-1 flying bomb launch ramp in Northern France. (The Cavalier collection)

offensive was relinquished and replaced by a joint RAF/USAAF planning organisation. Emphasis was once again on strategic effect, although support for the Army remained a subsidiary requirement, as demonstrated by strikes against the German redoubts at Boulogne and Calais, raids on rail centres during the Battle of Arnhem, and the breaching of the sea walls of Walcheren, enabling the Allies' use of the vital port of Antwerp.

Dissent continued over the most effective use of the bomber force. While the Air Ministry, supported by Portal, advocated the targeting of oil installations, Eisenhower's deputy, Air Marshal Tedder, prescribed attacks against communications. Both factions maintained that destruction of their target set would curtail reinforcements to the battlefront and thus bring the war to a quick conclusion. Harris, who considered both to be 'panacea' targets, stood firm in his belief that attacks on Germany's urban and industrial areas were the key to victory. A new directive issued on 25 September supported Portal and the oil protagonists, with transport and tank manufacture as second priority.

Effort, too, was diverted to address concerns of the Navy, notably the German battleship Tirpitz, hiding in northern Norway. Three attempts were made during September–November by a joint force of Lancasters from Nos 9 and 617 Squadrons utilising Barnes Wallis's 12,000lb Tallboy deep-penetration bomb, previously employed to destroy E- and U-boat pens and large V-weapons sites. The first of the demanding operations against *Tirpitz* required the use of a Russian airfield at

Yagodnik, near Archangel as a forward base. The others, launched from bases in Scotland, involved flights of over twelve hours' duration. *Tirpitz* was finally despatched on 12 November with three direct hits and several near misses. Denied the use of French bases, the U-boats were now operating from the Norwegian ports of Bergen and Trondheim. Main Force attacks against these ports in October to disrupt their operations met with limited success. Although vessels in harbour and dockside installations were hit, the U-boats in their concrete pens remained impervious to smaller-calibre bombs. As with the French ports, despite every effort to achieve accuracy, collateral damage to civilian property with the inevitable loss of life was inevitable.

Renewed support was required for the Allied troops as they neared the German border. In October Tallboy-carrying Lancasters breached the Kembs barrage on the upper Rhine in a low-level daylight attack to prevent the Germans releasing floodwaters against Allied troops crossing the river. At the beginning of December a series of operations with similar intent, by both Main Force and Tallboy aircraft, against the Urft Dam on the River Roer, were less successful, although the dams were captured before the Germans could destroy them.

Harris's frustration with official policy was assuaged in October with a directive intended to demonstrate the Allies' air superiority with a period of concentrated attacks; Operation Hurricane tasked Bomber Command to pursue objectives in the Ruhr and Rhineland. Cologne and Duisburg were twice attacked, the former also receiving the

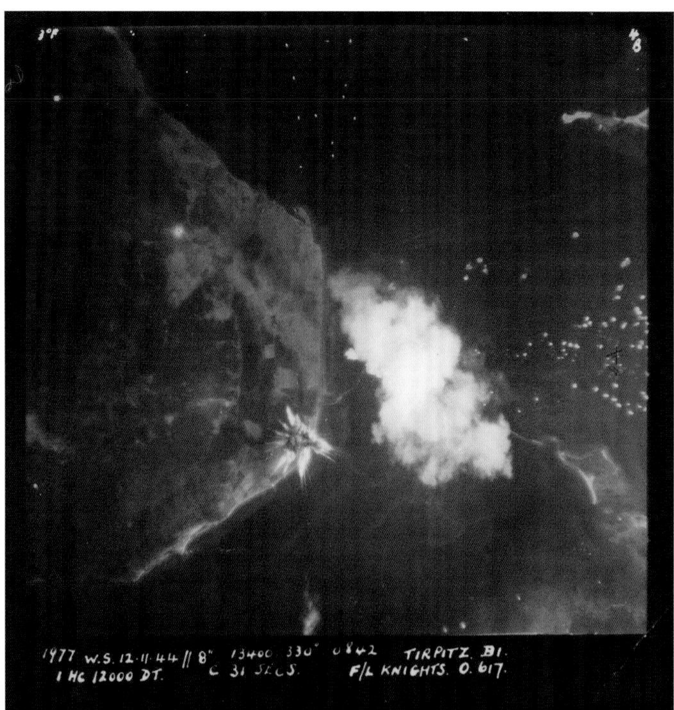

Left **A bombing photograph taken from Flight Lieutenant Knights' No. 617 Squadron Lancaster during the successful attack on the Tirpitz on 12 November 1944.** (The Twells collection)

Below **Two aircrew and three groundcrew in front of their No. 100 Group Mosquito. Each bomb painted on the nose of the Mosquito depicts a successful operation.** (The Franklin collection)

attentions of the USAAF by day. Essen and Brunswick were also the focus of raids. On 14/15 October Bomber Command aircraft took off on 1,576 sorties (1,294 against targets in Germany) – the greatest number in any one night – dropping 5,453 tons of explosives in total, 4,547 tons on Duisburg alone – again setting a new record. These figures would soon be exceeded on 16 November when the Command despatched 1,189 aircraft, releasing 5,689 tons on communications targets in support of the US First and Ninth Armies.

Despite inflicting severe damage, Operation Hurricane was short-lived and bombing effort reverted to oil production, but only 'when weather permitted'. By exploiting this caveat Harris was able to maintain his area offensive, to the extent that it brought serious approbation from Portal. However, after an acrimonious exchange of correspondence, the percentage of oil target sorties increased fourfold, but whenever the weather was marginal Harris favoured area targets. Nevertheless, during December 1944 Bomber Command hit more oil installations than the USAAF and in the following month dropped nearly twice the tonnage of bombs on oil targets than the Americans.

The Dortmund Ems and Mittelland Canals were both breached during this period and oil and transportation targets remained significant objectives for the remainder of the war, but the new year again saw official support for city targets. Instigated by the Chiefs of Staff and backed by Churchill, Operation Thunderclap tasked both Bomber Command and the USAAF to attack East German cities to support the Russian advance on the Eastern Front by disrupting transport and communications. The attacks would also undermine German morale and weaken resistance while demonstrating to friend and enemy alike the growing strength of the Western Allies' air power. The offensive was opened by the USAAF, with a raid on Berlin on 3 February. Nine days later Bomber Command attacked Dresden with 800 aircraft bombing in two waves. The attack resulted in a firestorm, the fires of which were rekindled by a strike by 300 aircraft of the US 8th Air Force on the 14th. That night over 700 Lancasters and Halifaxes were despatched to Chemnitz, another Thunderclap target. Although this raid was hampered by poor weather, on the 23/24th Pforzheim was subjected to intense attack resulting in severe damage and casualties. In addition, Mosquitoes of the Light Night Striking Force targeted Berlin for thirty-six consecutive nights from 20/21 February. The devastating potential of Bomber Command was now beyond all doubt. Now, at twice the strength of 1943, and with the German defences reduced, it was finally the weapon to which Harris had aspired in 1943. It was now able to destroy any small or medium town with one or two raids, as demonstrated when called upon to attack Goch, Kleve and Wesel, in support of ground forces advancing on the Rhine.

Meanwhile, attacks intensified against other familiar Ruhr and Rhineland industrial and communications centres, including Dortmund, Duisburg and Düsseldorf. On 12 March 1,108 aircraft took off for Dortmund, dropping 4,851 tons of explosives, the greatest number of planes despatched by Bomber Command to any one target in

one day during the entire war. Force size was not the only determinant of potency. Small numbers of Lancasters carrying Tallboy bombs and its larger stablemate, the 10-ton Grand Slam were used to great effect to sever key viaducts and bridges, thus isolating the Ruhr from its raw materials and preventing the transportation of finished armaments to the battle front.

On 1 April Churchill requested a review of the area offensive. With ground forces advancing into Germany and the final outcome only a matter of time, pressure was maintained on oil production, with added emphasis now on the U-boat construction yards at Hamburg, Bremen and Farge and elements of the German Fleet at Kiel and Swinemünde. Fighter activity waned as the success of the oil offensive starved the Luftwaffe of fuel, and although limited numbers of the new Me 262 jet fighters were encountered in the final months, effective though they were against daylight formations, they were too late to affect the final outcome.

Bomber Command's final large-scale attack against an urban target came on 14/15 April when 500 Lancasters and 12 Mosquitoes attacked Potsdam. The last time the 'heavies' had raided the Berlin area had been the 'night of the strong winds' when 72 of their number failed to return. Now with Allied air superiority assured, only one aircraft was lost. By now the strategic bomber offensive was coming to a close. The final sorties flown by the 'heavies' were on 25 April when 359 Lancasters attacked Hitler's southern redoubt at Berchtesgaden, while a mixed force of 482 Halifaxes and Lancasters attacked the coastal batteries on the island of Wangerooge. That night a smaller force of Lancasters mounted a strike on an oil refinery at Tønsberg in southern Norway. Each of these attacks was accompanied by Mosquito markers, and it was to be this versatile aircraft, rather than its heavier stablemates, which flew the Command's final bomber operation of the war on 2 May, when 179 aircraft, supported by Halifaxes, Fortresses and Liberators of No. 100 Group, attacked Kiel and targets in the surrounding area.

Even before the conclusion of the bombing war, Bomber Command was adapting to a more peaceful role. Commencing on 29 April and lasting until VE Day, 8 May, Operation Manna saw Lancasters and Mosquitoes flying 3,298 sorties to drop 6,672 tons of supplies and food to

Left An oblique photograph of the Duren railyards showing the widespread destruction following Allied bombing. (The Banks collection)

Right Starving Dutch civilians in western Holland express their appreciation to those flying an Avro Lancaster of Bomber Command, as they drop supplies of food as part of Operation Manna.

the starving population of western Holland. Operation Exodus, the repatriation of recently liberated British and Commonwealth prisoners of war, extended humanitarian concern, returning some 75,000 men, some of whom had not seen England for over five years.

Nevertheless, the global war was not yet over. In other theatres fighting continued and Bomber Command began the formation of 'Tiger Force' in preparation for operations against the Japanese in the Pacific theatre. By 15 August 1945, thirteen RAF and eight RCAF squadrons had been earmarked for re-equipment and training. Before it had the chance to deploy, the Pacific war was brought to a close by the dropping of two weapons, far exceeding the destructive power of anything in Bomber Command's arsenal. Tiger Force disbanded at the end of October 1945.

BROTHERS IN LIFE AND DEATH

FAILED TO RETURN

The Lincoln brothers, killed within weeks of one another while serving on the same squadron.

Avro Manchester L7470 waits for a 'green' from the Aldis lamp at the end of the runway at RAF Woolfox Lodge. It's the signal for the pilot, 21-year-old Eric Noble, to release the brakes, open the throttles to the two Rolls-Royce 'Vultures' and start his take-off run. By his side is the second pilot, another NCO, Lloyd Lincoln.

Lincoln, a 26-year-old Aucklander, has come a long way to join in the fight. Many thousands of miles and adventures have marked his life. He volunteered to join the Royal New Zealand Air Force (RNZAF) back in 1940, along with his younger brother, Stanley. Indeed, they have been allocated sequential service numbers: NZ41341 and NZ41342 respectively.

Their careers have mirrored one another's. Both qualified for pilot training and were posted to No. 2 Elementary Flying Training School (EFTS) at Bell Block. Lloyd soloed after 8.35 hours; Stanley some ten minutes sooner. The pair were then selected for multi-engined training and posted to No. 3 Service Flying Training School (SFTS) at Wigram. They were awarded their flying badges in May 1941.

Troopships were leaving at regular intervals, and the brothers boarded the Dominion Monarch in late July, arriving in the UK via Halifax, Nova Scotia, six weeks later. They were in good company; the ship carried some 466 aircrew including 183 trained pilots for service in the Fleet Air Arm and RAF.

In the UK their paths diverge, but they are soon reunited, both men being posted to No. 61 Squadron. Both will start as second pilots before being made captain of their own aircraft. Both will be flying the Avro Manchester, a rather disappointing aircraft whose engines are to prove woefully unreliable.

Stanley is not flying on the night of 6 April 1942; he's flown five trips in March, including a couple of 'gardening' sorties, sowing mines. Lloyd's name, however, is on the Battle Order and it's a tough target: Essen in the Ruhr, the heartland of Germany's war machine. He's been there before and been 'coned', trapped in the sinister beam of a searchlight. He knows how dangerous it can be. The squadron is contributing five crews to an attacking force of more than 150, a mixed bag of Hampdens, Wellingtons, Stirlings and Manchesters. Two Manchesters return to Woolfox Lodge with engine troubles. A third does not return at all. Noble, Lincoln, and all of their crew – including Sergeants Robert Newton, Kenneth Leyshon, Donavan Kent, Neville Patton and George Walker – are missing. They have been shot down by a German night fighter and all are dead.

Despite his grievous loss, Stanley must go on flying. Not long after his brother's death, the squadron converts from two engines to four, their Manchesters being replaced by Lancasters. It also moves base, to Syerston. Stanley flies his first Lancaster sortie on 29 May, a 'gardening' trip that passes without incident. The next night and the squadron contributes seventeen Lancasters to a maximum show, the first of the 'thousand-bomber' raids. Stanley knows it's a big one, as both flight commanders – Squadron Leaders Rupert Gascoyne-Cecil and Peter West – are in the air. Cologne is besieged. Lincoln's aircraft is 'coned' for a terrifying few minutes and hit by flak, but makes it home in one piece.

On 2 June he, too, is detailed to attack Essen. The raid is a modest affair. Stanley is second pilot in R5613. His regular captain, 21-year-old Pilot Officer Ralph Clark, used to be an NCO. He holds the Distinguished Flying Medal (DFM) from a tour with No. 144 Squadron in 1941. The rest of the crew are also all NCOs: Edward Patchett, Oliver Beswick, William Griffiths and Norman Hartley, as well as another DFM recipient, Alastair McKelvie.

One minute after midnight, their aircraft heads out into the night. Of the five crews from No. 61 Squadron taking part, two fail to return, including R5613, the victim of a night fighter. Six of the seven men are killed. The seventh, William Griffiths, is on the run, and will eventually reach safety in Spain.

Another telegram is soon on its way to Mount Albert in Auckland where a father will suffer the loss of his second son in only a few short weeks, and whose heart will break, never to recover.

THE INTERNATIONAL BOMBER COMMAND CENTRE
PART TWO THE LOSSES DATABASE

Left Veterans, from left to right, Jo Lancaster, Alan McDonald, Henry Wagner, Gordon Mellor, Ken Johnson, Jack Smith, Gerry Norwood, and Hal Gardner at the Spire, seated amidst the walls that commemorate all those lost serving with Bomber Command.

From the very start of the project, one of the key aims was to be the international centre for information on the nearly 58,000 aircrew and others members of Bomber Command who lost their lives. Plenty of research had, of course, been done already, but it had never been free to use, in digital format, and put together with such depth of information. This made the task for the team all the more challenging in terms of sourcing, verifying, managing and cross-referencing the information.

The losses database began at the launch event in East Kirkby in Lincolnshire. The winning design for the monument was to be a Spire made up of the names of the 26,000 Bomber Command losses that were contained in the three books of remembrance at Lincoln Cathedral – the rolls of honour. Dave Gilbert, who attended the event, wondered how the names would be transferred on to the structure, to which the answer was 'I've no idea – have you?' And so began a journey that would involve international cooperation, investigation and the collation of data on a never-before-done scale.

From the start it was clear that if the data was to be of any use to families, researchers and future generations, the names needed to be stored on a computer, and a spreadsheet was designed to hold the mass of information.

The first challenge the team faced was how to transfer the names from the three rolls of honour in Lincoln Cathedral to the computer, so enquiries were made to the RAF who own the books. They said that if the books were removed, they would not only need huge levels of insurance, but would also be transported in specially built containers because of their significant importance and value. Such criteria were out of the reach of the Trust, which was in its infancy, so a project was started to photograph the books (without a flash to protect the pages), and transcribe the names and other information contained in them (which covered Nos 1 and 5 Groups and the OTUs) into the spreadsheet 'by hand'.

It was a laborious process but one which was at the core of one of the key elements of the whole project. Some of the information was incomplete – various names also had their service number, others did not, while many had squadron details, but some not. In fact, the only element they all did have in common was the date of death. The information was sketchy at best, but it was a start.

With the initial transcription exercise complete, and as names were input, the real human story of Bomber Command became clearer. Accounts of four sets of brothers dying on the same day, and other siblings within days of each other, all brought home the importance of recording their names so they would not be forgotten.

Whenever there was a query, more 'digging' needed to be done to resolve the issue – gathering more information and finding new sources that could enrich the main database.

With all the names transcribed, they were checked against the original rolls of honour to ensure that they had been correctly input. But while these were compared with the original data, the exercise did not of course ensure that the details on the rolls were right – a particularly important aspect as they were to be used as the source for some very important and expensive metalwork in the near future: the construction of the memorial walls.

A random sample of 500 names was chosen and checked against the Commonwealth War Graves Commission (CWGC) website – manually at first. This threw up a number of inconsistencies – in fact about fifteen per cent of the names came back with some form of error, such as wrong initials, incorrect spelling of surnames etc. With the number of panels planned for the memorial, the law of averages said that every panel at the memorial would have at least one mistake on it – an unacceptable position to be in and one which the Trust was not prepared to live with. Each name represented an individual and the least they felt they could do was get their details correct. The big issue at this stage was that

Left Ensuring that the modern generation remember those who made the ultimate sacrifice with Bomber Command.

Below One of the stones in the Ribbon of Remembrance, which frames the Memorial Avenue that connects the Chadwick Centre to the Spire Memorial and Walls of Names.

the deadline for handing over the names to the architects was looming large, which meant a solution needed to be found fast.

A much larger task than a single man could undertake, the local radio station, BBC Radio Lincolnshire, was contacted. They were very good and invited Dave into the studios to explain more about the project, what its aims were and his particular problem with regard to checking the names. After the interview, it's fair to say that the switchboards 'lit up' as never before with people offering to help. Forty calls were received in the first half-hour and ninety volunteers were signed up in the course of the first week – a remarkable response that showed the strength of feeling about the project within Lincolnshire and beyond. In fact, this was the real start of the volunteer process, and the engagement of the county and people across the UK in the scheme.

The first 26,000 names were divided between the number of volunteers, which gave each volunteer a batch of around 300 names to check. The purpose of the exercise at this stage was really to ensure that just the names were correct. However, in order to do that the volunteers needed to look on the CWGC website. Obviously the CWGC website records only the 'Commonwealth' war graves, so it excluded the French, Americans, Norwegians etc., but in ninety-five per cent of cases the names could be found and verified.

It was then that the real 'sea change' in the size and depth of the losses database happened. The CWGC site contained much additional information that was not at that

Far left The unveiling of the Spire, 2 October 2015.

Left The Queen's Ninetieth birthday beacon ceremony, 21 April 2016.

time held in the losses database, such as cemetery details, grave references and next of kin. If the losses database was to be comprehensive, it made absolute sense to take the additional information held on the CWGC site and bring it back into the IBCC losses database. Instructions were given to volunteers via video and notes that while the number one task was to make sure the information on the losses database was accurate, while they were at the CWGC site, they should bring the additional information across to enrich the losses database. It was at this stage that the database started the journey from simply being a list, to a comprehensive information resource for use across the world. By this time also there were plans in place to have the losses on the IBCC website and searchable by relatives and researchers, so the gathering of additional information became a priority.

And so the mammoth task of training and coordinating volunteers across the world began, and the information they gathered started to be assimilated into the database.

But there was more to be done. The CWGC website contained the information it was intended to – grave locations, names etc., but of course it held no operational details – what operations the individuals died on, what aircraft they were flying, who they were flying with, what circumstances led to their death etc., and so the next phase of the IBCC project began.

This is where the 'Chorley books' came into the project. The books, by W.R. Chorley, have been used by military historians for decades, and they catalogue the aircraft losses suffered by Bomber Command. They contain the majority of the losses of aircraft including the crew and operational details. The books are, however, by their nature based on the aircraft and not the crew. For example, they would not list a situation where a rear gunner died on an operation, but the aircraft came back to base. But it was undoubtedly a massive boost to the project when Bill Chorley readily gave his permission to use the data he had accumulated over the years to be combined and cross-referred into the losses database.

Phase 2 (the remaining 32,000 names) was the last opportunity to make sure the project had all the names, so visits were made to Ely and York Cathedrals to photograph their rolls of honour. Contact was also made with Nanton, Alberta (home of the Canadian Bomber Command Memorial), as well as perusing POW lists to account for those who died in captivity, squadron association lists and websites, other books and information resources. The need was for detail and the 'Middlebrook Book' (Bomber Command War Diaries by Martin Middlebrook and Chris Everitt) gave much more operational information, providing real depth to the individual stories, including details of how many aircraft were involved in the operations, what mix of aircraft they were, what percentage didn't return etc. All valuable information for cross-checking, data verification and enrichment.

One of the problems this raised was the multiple instances of the same name and other repeats, but of course this generated further work in identifying which was correct – such an important aspect of the project.

The scale of the operation was becoming clear, so

bespoke programmes were written in conjunction with the CWGC to take information from their site and feed it into the losses database. This 'automation' speeded up the information-gathering process and helped greatly with the data verification.

At the end of the day the database deals not with data, but with people, and as the team entered the data, the real human aspect of the work became clearer. The last column on the spreadsheet is the date of death and the finality of this constantly focused the mind on the real issue and meaning of the task. The research uncovered many heroic stories, deserving of, but not receiving, recognition and honour, which again brought sharply into focus the true sacrifice of such brave individuals.

As extensive and comprehensive the losses database is, it needs to be something more than just a factual content. It needs humanising with information about the people – their careers before they joined the Service, what they gave up to enlist, what type of person they were etc. The team hope some of this knowledge will come from family members researching their family history, friends searching the database, or one of many other sources. With Phase 1 being establishing the first 26,300 names on the memorial, and Phase 2 the next 32,000, what lies ahead for the losses database? Phase 3 will involve looking at pre- and post-war deaths (Bomber Command was active between 1936 and 1968), and, just as importantly, investigating those occurring in the Mediterranean and Middle East, of which there are around 5,000, but which suffer from a real lack of data and documented information that will make the task more than a little challenging. This will involve trips by the team to the National Archives at Kew to check the operational record books, as will filling in the gaps on Phases 1 and 2, where there remain some individuals with very little operational or personal details.

Importantly, the project team are determined to include all Bomber Command losses – in particular ground crew who played such a significant part in their operations and civilians who perished on the planes, such as war reporters (three of whom died on the same night), and those on training exercises. In many cases these are not commemorated anywhere else, but with the losses database project, there will now be a permanent home for their memories.

The losses database represents an immense effort on an internationally significant programme, that takes information beyond being just data to generating memories of people who gave their lives and who will never be forgotten.

Right **The Battle of Britain Memorial Flight's Avro Lancaster salutes the IBCC's Memorial Spire.**

Donald Nicholson, a 93-year-old Bomber Command veteran discovered he had lost his medals at the unveiling on 2 October 2015. Donald was taken to the onsite Media Centre, and a national appeal to find the medals was launched. At the same time the IBCC posted an appeal on its social media sites. On 6 October the IBCC asked if anyone could come and help conduct a search with metal detectors on the Friday afternoon. The response to the appeal to search for the medals was unbelievable, with 85 detectorists arriving at the site. Unfortunately, the medals were not located but during the search two pieces of a Lancaster were found. In 1942 two Lancasters from Waddington had crashed, in bad visibility, over the field next to the IBCC. The IBCC contacted Donald to tell him that the medals had not been found but he was happy to report that the medals had been handed in at his local Police station. As the medals were unnamed the Police had not been able to return them until a Sergeant saw the media coverage of the appeal and made contact. One thing was left outstanding. Donald had not been able to wear his medals at the event. Working with BBC Radio Lincolnshire the IBCC brought him back on the 14 October and a brief ceremony

was held including a flypast of the BBMF Lancaster, Donald commenting 'I'm over the moon and on top of the bloody world!' Donald was also able to lay poppies in the names of the crew he had flown with but were killed in 1943. He said, 'I have finally been able to close the book on their deaths. It's taken me 72 years to be able to do that. Thank you for making it possible.' (Sadly Donald passed away in March 2016.)

OLD PILOTS AND BOLD PILOTS

4

Harry Ashworth and crew. Three did not make it out of their burning bomber.

At forty years of age, Squadron Leader Harold Ashworth is rather old for a bomber pilot. But he has a huge amount of flying experience. As a civilian pilot, he'd competed in the King's Cup Air Race in the 1920s and joined the RAF at the outbreak of war. Now he is the popular 'B' Flight commander with No. 218 Squadron at Marham.

His crew are of mixed experience and class: the second pilot, Shanks, the navigator, Green, and the air bomber, Attwood, are all commissioned men. Alan Green has flown a dozen or so trips with another pilot. The flight engineer, Hayden, the wireless operator, Watt, and the two air gunners, Whitehead and Mulroy, are sergeants.

By the winter of 1941–42 No. 218 Squadron has converted from the faithful Wellington to the mighty four-engined Stirling. They also have a new CO, Wing Commander Paul Holder, DFC, a South African by birth. He is a veteran of the defence of Habbaniya in Iraq. Most recently the squadron has been attacking the German Navy's capital ships, Scharnhorst and Gneisenau, as they make their infamous 'Channel dash'.

Targets are many and varied. A strategic pattern has not yet emerged, although Essen in the Ruhr is proving a popular choice alongside specialist targets such as the Skoda works at Pilsen. The latter is part of a 'secret' war, with its own code-name – Operation Canonbury – liaising with Czech resistance fighters on the ground.

On 4 May 1942 Ashworth and his crew are detailed for a solo Nickel raid on Laon, dropping two tons of propaganda while the remainder of the squadron is split between Pilsen (again) and a raid on Stuttgart. On their return, and in the vicinity of Norwich, they suddenly come under attack, an early victim of 'friendly fire'. Their assailant is an RAF Hurricane, working in tandem with a twin-engined Havoc, and it has found its mark, striking the Stirling (Q9313) a mortal blow. The skipper gives the order to bail out and the crew take to their 'chutes. The pilot goes to great lengths to ensure everyone has gone – an action that later earns him a Distinguished Flying Cross – before he himself takes a leap of faith into the night. Fortunately, no one is hurt and all become instant members of the Caterpillar Club, their lives saved by an Irvin parachute.

It is not their only brush with danger. On 1 June 1942 they return to Essen for one of the showpiece 'thousand-bomber' raids. They are attacked by a single-engined fighter just as their Stirling is on its final approach. A short burst from only fifty yards misses its target, and the fighter disappears into the night. A few nights later and the eight men are joined by a ninth, Group Captain Andrew 'Square' McKee, DSO, DFC, AFC, a New Zealander, for another attack on Bremen. Thankfully the night passes without incident. It does not do well to lose the station commander.

On 20 June they are one of nine squadron aircraft briefed for an attack on Emden. They are flying Stirling W7530 Q-Queenie with a new second pilot, Desmond Plunkett. He is a former instructor but this is only his third trip. As they leave the target area, a night fighter moves in to attack. Canon shells rip through a fuel tank, setting the aircraft on fire. Ashworth desperately fights with the controls as the aircraft becomes increasingly unstable. It is useless, and he gives the order to bail out.

The aircraft is heard and seen from the ground, on fire and spiralling in ever decreasing circles. A parachute blossoms as the first of the five aircrew who will survive drifts to the ground. Three don't make it: Harold Ashworth, William Watt and William Whitehead, the two NCOs being only half their captain's age. The rest are ultimately taken prisoner, where their adventures begin anew.

1

2

3

4

5

The International Bomber Command Centre

BOMBER COMMAND
PART ONE THE AIRCREW STORY

In his poem 'Six Young Men', inspired by a photograph taken at Lumbs Falls near Hebden Bridge, the late poet Laureate Ted Hughes opens with 'The celluloid of a photograph holds them well', and he goes on to describe the ageing of the photograph but not that of the youthful faces of the men in it. The little idiosyncrasies he perceives give humanity to the men, who are about to enter the maelstrom of the First World War. Their humanity does not last long: 'Six months after this picture they were all dead.' The men –'all were killed' – no longer exist, but the setting of the photograph remains. The photograph itself exists. The very writing of the poem has given permanence to the photograph and the image. But still 'all were killed'.

These men were from the generation whose sons, nephews and cousins would have to go to war again two decades after the first global conflict of the twentieth century. The shadow of the horrors of the Somme and Passchendaele remained. But perhaps the war that broke out in September 1939 would be different. Throughout the Second World War the faces of the new generation of servicemen were captured by official and newspaper photographers, and by the cameras of the individual men and women themselves: two friends larking around, a wedding picture, an aircrew in front of their aircraft. These moments in time could be used for propaganda and official records. They could be included with letters sent to and from a Royal Air Force station. They may have turned out to be pictures of men in civilian clothes for use in false evasion documents, or perhaps a snap of a downed airman taken by a German photographer for use on a prisoner-of-war identity card. Each image displayed the little idiosyncrasies of this generation's humanity. After the war, official archives and museums preserved many of these images, but many more were held in personal archives, which historians and authors may occasionally have the privilege of seeing. Thousands may be lost at the final passing of the wartime generation (although the extraordinary dedication of all those at the International Bomber Command Centre's digital archive are doing their best to find and preserve these images).

The geographical spaces and locations where these photographs were taken are still there of course, but the backdrops have changed over time, and perhaps the details no longer remain. Maybe there are some derelict or crumbled airfield buildings hosting the ghosts of times past. But what happened to the men in the pictures in the months after they were originally taken? Did death come too soon in their young lives? They may have avoided that fate, for the time being, but perhaps someone they knew was not so fortunate. Why and how did they find themselves in these pictures? How many of these faces did not 'grow old' as those that were left had, as expressed in Laurence Binyon's poem 'For the Fallen'? How many had not been wearied by age or by the years condemned? How many of these celluloid expressions, smiles, indifferences, grimaces, would, sooner than hoped for, become a memory to those they knew?

What was it that motivated a young man to take to the air in a tube protected by a thin sheet of metal, wood, or fabric, and face extremes of weather, knowing they would be hunted down by an extremely hostile adversary who was determined to shoot them from the sky? During the years 1939 to 1945, 125,000 young men would decide to do just that – take to the air with RAF Bomber Command – a new generation of young men whose forefathers had been of the generation who had fought in the First World War. The then modern youth did not want to go through what their predecessors had experienced. Tony Iveson served as both a Spitfire pilot during the Battle of Britain and as a Lancaster pilot with Bomber Command's No. 617 Squadron: 'My father was wounded on the first day of the Somme, 1 July 1916, and bore a terrible battle scar on his chest. One knew a lot about trench warfare.' Perhaps these new recruits had been inspired by the sheer wonder of what was still a relatively new technology – powered aerial flight. Pilot Tiny Cooling served with No. 9 Squadron

and recalled during the early days of his training: 'It was a beautiful sunny day and I was diving down into these cumulus valleys, shrieking with delight. It was bloody marvellous.' Then there were the stories they could read of the aviators of the First World War and perhaps they had seen and been enthralled by the thrilling exploits of Sir Alan Cobham and his Flying Circus in the early part of the 1930s, which toured the country entertaining the crowds. Jo Lancaster, who would go on to fly Wellingtons and Lancasters, had his first flight with the 'Circus'. 'It cost five shillings. I remember it was an Avro 504. I sat in the rear cockpit which had a 'fore and aft' bench-type seat and sat astride this. My fellow passenger was a girl, unknown to me, who just kept her head down and squealed through-

Previous page, main image
1. The typical trades of a Bomber Command crew, from left to right, Arthur Darlow (pilot), F/O Constable (navigator), Alec Nethery (bomb aimer), Allan Burrell (wireless operator), Philip Richards (flight engineer), Trevor Utton (mid upper gunner), Don Copeland (rear gunner).

2. Dennis Moore, second from the left, 'Taken at No. 15 Squadron shortly after moving from No. 218 Squadron'. (The Moore collection)

3. Tom Jones, second from the right and his No. 7 Squadron crew. (The Jones collection)

4. From the Smith collection, a group of pilots in front of a Handley Page Hereford at No. 16 Operatonal Training Unit.

5. Pilot Hedley Madgett, standing first left. Hedley, along with his 61 Squadron crew, were all killed on the 17/18 August 1943 Peenemünde raid. (The Madgett collection)

6. Navigator Roy Smith, standing second from the right, with his air-crew and groundcrew colleagues in front of their Short Stirling.

Opposite Veteran Wellington and Lancaster pilot Jo Lancaster DFC.

Above Veteran Halifax and Mosquito pilot George Dunn DFC L d'H.

Right Standing second from right, George Dunn of the Home Guard.

out the trip.' Lancaster and Mosquito pilot Benny Goodman recalled: 'With visions of the trenches I decided that I would like to fly. I had read every book that I could find on the aces of the First World War. My father took me to see a display by Alan Cobham and his flying circus. I greatly admired these people tearing around in their flying machines and this really triggered me off.'

The youth that went on to serve with Bomber Command were motivated to join the RAF for the thrill of flying, to break from the mundane, and to avoid the horrors of the previous generation. This is a common theme. There was no sadistic motivation – no desire to kill. This group of educated young men were aware of the political situation and, once the war started, what many of them witnessed hardened their resolve to defend their country, their way of life. As Dam Buster Johnny Johnson recalled: 'We were in a pickle. ... My motivation was anti Hitler.'

George Dunn, a future Halifax and Mosquito pilot, was only a couple of weeks short of his seventeenth birthday when the war broke out. Subsequently, and keen to do his bit, George heard on the wireless that able-bodied men over the age of seventeen were required to become Local Defence Volunteers. George signed up: 'All we had was an armband with LDV on, no weapons.'

"I DON'T KNOW WHAT WE WOULD HAVE DONE IF ANY PARACHUTISTS HAD DROPPED DOWN FROM THE SKY. AFTER A WHILE WE GOT UNIFORMS, AND A RIFLE, AND BECAME KNOWN AS THE HOME GUARD. JUST AFTER MY

18TH BIRTHDAY I DECIDED I COULDN'T WAIT FOR CALL-UP. I DIDN'T FANCY THE ARMY, AND I HAD A FEAR OF WATER – I DIDN'T FANCY DROWNING. LIVING ON THE NORTH KENT COAST I HAD SEEN A LOT OF THE BATTLE OF BRITAIN, AND WHEN THE LONDON BLITZ STARTED THE GERMAN BOMBERS, HORDES OF THEM, CAME UP THE THAMES ESTUARY. SO I DECIDED I WOULD GO UP TO CHATHAM AND VOLUNTEER FOR AIRCREW." GEORGE DUNN

Hal Gardner served as a navigator with Nos 106 and 189 Squadrons. At the start of the war he was living and working in Brighton: 'I was 17 and saw an awful lot of the Battle of Britain.' Hal was also part of his local Air Training Corps (ATC) squadron. Originally called the Air Defence Cadet Corps, founded in 1938 to encourage young people to begin aviation training at an early age, the renamed ATC would recruit tens of thousands of young cadets, many of whom would go on to serve in Bomber Command. Hal was certainly one of those who was enthused: 'We all wanted to be Spitfire pilots. That's what we aimed for, but it didn't always work that way.'

The Royal Air Force were keen to tempt these boys into this region of almost supernatural wonder. The recruitment posters sold the thrill of flying, adventure, a clean war and being part of a team – the ATC, a crew, a squadron. Young men were to be enticed away from the world of the common day; it was an opportunity for adventure.

When the war broke out Gerry Norwood, who would become a Lancaster air gunner, was in a reserved

Left Veteran Lancaster navigator Hal Gardner.

Below Veteran Lancaster air gunner Gerry Norwood.

Below right Gerry Norwood, third from the left, in front of 'Queenie!'.

occupation. He was certainly keen to break free from the mundane ways of everyday life. At Gerry's place of work, 'all the males worked on 12 hour shifts – females on the day shift and males on the night shift, 7 nights a week, 12 hours a night, 8am to 8pm'.

"YOU NEVER GOT A DAY OFF. I NOT ONLY GOT FED UP WITH IT BUT ALL MY MATES HAD JOINED UP IN THE FORCES. I FELT AS IF I WASN'T DOING ENOUGH. I SAID TO ONE OF THE FOREMEN, 'I'M GOING TO JOIN UP.' HE SAID, 'YOU CAN'T. YOU'RE A RESERVED OCCUPATION.' I REPLIED, 'WELL I'M GOING TO TRY.' WHEN I CAME OFF WORK ONE FRIDAY MORNING I WENT STRAIGHT UP TO EDGWARE, LONDON, AND THE DEANSBROOK ROAD DRILL HALL. I WENT TO THE ARMY BLOKE AND SAID I WANTED TO VOLUNTEER. HE CHECKED AND REPLIED, 'NO. SORRY, RESERVED OCCUPATION. TRY THE NAVY.' I TRIED THE NAVY. SAME THING, 'THERE'S NO WAY WE CAN TAKE YOU FOR ANYTHING.' THEN I WENT FOR THE AIR FORCE AND HE SAID, 'WE CAN'T TAKE YOU BUT IF YOU ARE WILLING TO FLY WE CAN TAKE YOU FOR AIRCREW. BUT YOU HAVEN'T GOT MUCH HOPE BECAUSE YOU HAVEN'T HAD THE RIGHT EDUCATION.' I TOLD HIM TO PUT ME DOWN AND I'D TRY. ABOUT 5 WEEKS LATER I GOT A TRAVEL WARRANT AND A LETTER FROM THE AIR FORCE TO REPORT TO ONE OF THE COLLEGES AT OXFORD."
GERRY NORWOOD

When he arrived in Oxford Gerry sat some exams and was put through numerous medical and aptitude tests.

"I WAS CALLED BEFORE THE SELECTION BOARD AND WAS TOLD, 'WE ARE EXTREMELY FASCINATED THAT YOU HAVE PASSED EVERYTHING EXCEPT MATHEMATICS. WITH YOUR EDUCATION WE CANNOT SEE HOW YOU CAN GO THROUGH. BUT IF YOU ARE WILLING TO GO TO NIGHT SCHOOL AND DO MATHEMATICS WE'LL TAKE YOU INTO THE AIR FORCE.' I SAID I WAS WILLING AND THEY SWORE ME IN THAT DAY, GAVE ME THE KING'S SHILLING AND MY NUMBER 1604811. I WENT BACK TO WORK AND THEY SAID, 'IT'S IMPOSSIBLE.' I REPLIED 'NO IT'S NOT. I'VE JOINED AND THAT'S IT.'"
GERRY NORWOOD

Having volunteered for aircrew duties with the Royal Air Force, and been called up, the next months would be spent learning the basics, before taking up one of the various trades. For some their training was in the UK, but many men, specific to other required roles, were shipped out to friendly countries as part of the British Commonwealth Air Training Plan, away from hostile and cramped airspace, to learn to fly, to navigate, to drop bombs. George Dunn, destined to be a pilot, was one of thousands of aspiring aviators who would be sent across the seas to develop his skills. 'It was Christmas 1941 and I found myself on draft for Canada.'

"WE ARRIVED IN HALIFAX, NOVA SCOTIA, AND THEN WENT BY RAIL TO A TRANSIT CAMP AT MONCTON. WHEN WE HAD OUR MEAL THAT NIGHT IT WAS MARVELLOUS. THE TABLE WAS COVERED WITH STEAK, BACON, EGGS, ICE CREAM, JAM. WE THOUGHT THIS IS GOING TO BE GREAT – NINE MONTHS OF THIS AND WE ARE REALLY GOING TO ENJOY IT. BUT WHEN WE GOT OUT TO SASKATCHEWAN WE FOUND

McMASTER MS
McMILLAN CO • McMILLAN D
McMILLAN M • McMILLAN W • McMILLAN WH
McMINN RJ • McMONAGLE LR • McMORRAN AE • McMORRAN GMS
McMULLAN WN • McMULLIN CAB • McMULLIN H
McMURCHY LS • McMURCHY WA • McMURDO JW • McMURTRIE JC • McNAB DC
McNABB RH • McNAIR A • McNAIR BR • McNAIR NW • McNAIR WA
McNALLY AF • McNALLY JJ • McNAMARA LGL • McNAMARA MJ
McNAMARA R • McNAMEE WA • McNAUGHT DH • McNAUGHT WJB
McNAUGHTON IGA • McNAUGHTON WR • McNAY IR • McNEARY J • McNEIL A
McNEIL DS • McNEIL F • McNEIL MG • McNEILL A • McNEILL D • McNEILL JF
McNEILL RF • McNEILL RM • McNEILL ST • McNEILL TH • McNEILLY WS
McNICHOL GA • McNICOL A
McPHAIL AM • McPHAIL DH
McPHEE JA • McPHEE JN
McPHERSON RD • McPHIE KC
McQUADE W • McQUADE WJ
McQUEEN WS • McQUILLAN MR
McRITCHIE RE • McRUER DW

Left Veteran Wellington and Halifax navigator Gordon Mellor looking on to the name of his air gunner Sergeant Norman Tolson McMaster who lost his life when their Halifax was shot down on the night of 5/6 October 1942.

Below Air gunner Steve Bethell.

OUT WE WERE ON ROYAL AIR FORCE RATIONS, AND NOT ROYAL CANADIAN AIR FORCE RATIONS! AT THAT TIME SASKATCHEWAN WAS COVERED IN SNOW. ALL MY ELEMENTARY TRAINING ON TIGER MOTHS AND SUBSEQUENTLY ON TWIN-ENGINED ANSONS WAS DONE IN SNOWY CONDITIONS, SOMETIMES REAL BLIZZARDS. EVENTUALLY I CAME BACK TO THE UK IN SEPTEMBER 1942 AND DID A COURSE ON OXFORDS TO GET USED TO FLYING IN THIS COUNTRY. CANADA WAS WELL LIT UP WHEREAS THIS COUNTRY WAS IN BLACKOUT." GEORGE DUNN

Air gunner Steve Bethell signed up in July 1942 and was called up two months later, proceeding to Air Crew Reception Centre at Lord's cricket ground, London:

"WE WERE KITTED OUT ON THE CRICKET GROUND AND WE USED TO EAT IN LONDON ZOO – ALL MARCHED DOWN THERE, VERY WELL ORGANISED. FROM LORD'S I WAS SENT TO THE INITIAL TRAINING WING [ITW] AT BRIDLINGTON IN YORKSHIRE WHERE WE WERE PUT IN PRIVATE LODGINGS, CROWDED OF COURSE. WE SPENT ABOUT 12 WEEKS THERE AND WERE NEVER SMARTER THAN WHEN WE'D FINISHED. IT WAS ALL DRILL AND SUCH LIKE, RUNNING ROUND THE BEACHES. I FELT VERY FIT. WITH AIRCREW ONCE YOU'D FINISHED ITW THEN AS FAR AS ALL THAT PHYSICAL WORK WAS CONCERNED IT WAS ALL DOWNHILL." STEVE BETHELL

Dave Fellowes, a Lancaster rear gunner, volunteered for the Pilot, Navigator and Bomb Aimer scheme. 'I went down to Lord's cricket ground, Air Crew Reception Centre, St John's Wood, and while there we were issued with uniforms, inoculated, tested for one thing, tested for another, could you hear properly, were you colour blind. A hundred and one different things. Subsequently, and via ITW and a Grading School at Longtown, Carlisle, Dave ended up at Heaton Park, Manchester:

"NOT A VERY NICE PLACE. HORRIBLE CAMP. LOT OF AWFUL CORPORALS RUNNING AROUND GIVING YOU A LOT OF STICK. THERE WAS A HOLD UP OF PEOPLE GOING OFF TO SOUTH AFRICA OR CANADA TO GET TRAINING. I THOUGHT TO MYSELF, 'I'M NOT GOING TO GET IN TO THIS WAR.' A NOTICE WENT UP AND IT SAID THAT YOU COULD BE AN AIR GUNNER IN THREE WEEKS. I WENT AND SAID, 'I'LL DO THIS!' I WANTED TO GET INTO A BIT OF ACTION. SIMPLY THAT AND IT WAS THE BEST WAY I COULD DO IT. IT WAS WHAT WAS EXPECTED OF US. I HAD MADE MY MIND UP AND I WASN'T GOING IN THE ARMY AND I CERTAINLY WASN'T GOING IN THE NAVY AND I CERTAINLY WASN'T GOING TO GO DOWN THE MINES. SO THAT WAS MY BEST OPTION AND NO REGRETS." DAVE FELLOWES

With their trade allocated and respective brevets issued, the next phase was for the airmen to become part of a crew, which, as the war progressed, became a somewhat random procedure, although it could throw up extraordinary coincidences, and seemed remarkably effective, usually leading to lifelong bonds of friendship.

Following some time at home on leave, Dave Fellowes had received a letter and a travel warrant telling him to report to the Operational Training Unit (OTU) at Hixon,

Staffordshire. Bomber Command was very much a multi-national force, with personnel from other British Commonwealth nations, who had answered the call from what many still viewed as the 'mother country', or from nations in which Nazi oppression had forced the men to find sanctuary in the United Kingdom where they could still fight back. For Dave Fellowes his destiny was with one particular group of airmen from an Allied nation. Dave had travelled to Crewe train station 'and changed there for Stafford':

"I WAS IN THIS EMPTY COMPARTMENT AND ALL OF A SUDDEN THREE AUSTRALIAN FLIGHT SERGEANT PILOTS CAME IN. 'GOING TO HIXON MATE?' 'YES', I REPLIED. I GOT TALKING TO THEM AND SAID TO ONE, 'WHERE DO YOU COME FROM?' HE SAID 'SYDNEY.' 'OH', I SAID, 'DO YOU KNOW MARRICKVILLE?' HE SAID 'YEH, I LIVE IN MARRICKVILLE. WHY'S THAT?' I TOLD HIM I HAD AN AUNT WHO LIVES THERE; SHE HAD GONE AFTER THE FIRST WORLD WAR. HE ASKED WHAT THE ADDRESS WAS, WHICH I GAVE HIM. 'WHAT'S HER NAME?' HE ASKED. 'MRS EVANS,' I REPLIED. 'OH MATE. IT'S MY MUM'S BEST CHAPEL FRIEND.' SO I SAID, 'SHALL WE CREW UP NOW?' 'YES, WHY NOT.'"
DAVE FELLOWES

At the OTU Dave and his Australian pilot Arthur Whitmarsh formed up with the rest of their crew.

"WE WERE PUT IN A BIG ROOM AND IT WAS 'SORT YOURSELVES OUT. WE WANT SIX OF YOU.' WE WERE LOOKING FOR A NAVIGATOR AND SAID HE'D GOT TO BE A STUDIOUS LOOKING INDIVIDUAL. I FOUND ANOTHER GUNNER WHO WAS AT GUNNERY SCHOOL WITH ME. THE WIRELESS OPERATOR WAS ANOTHER AUSTRALIAN, GOOD LAD, THE BOMB AIMER CAME FROM SCOTLAND, AND THE NAVIGATOR WE FOUND WAS A BLOODY GOOD TROMBONE PLAYER.

WHEN WE FINISHED AT OTU WE ALL WENT ON LEAVE AND THEN WERE SENT UP TO LINDHOLME AND NO. 1656 HEAVY CONVERSION UNIT (HCU). WHILE WE WERE THERE, THERE WAS A LITTLE BIT OF DELAY AND WE WERE IN A CAMP CALLED BOSTON PARK. WE WERE HAVING FIGHTER AFFILIATION EXERCISES ON OUR BICYCLES – ILLEGITIMATELY – MESSING ABOUT PLAYING FIGHTERS. THE STATION WARRANT OFFICER CAUGHT US AND READ THE RIOT ACT.

AT HCU WE WERE FLYING HALIFAXES AND DID A COUPLE OF 'SPECIALS', ONE OF WHICH WAS A RUN UP OVER THE NORTH SEA TO HOLLAND. A 'SPOOF' DIVERTING ATTENTION FROM THE MAIN STREAM. WHEN WE GOT BACK WE WERE DIVERTED AND THE GROUND STAFF ENGINEER CAME ROUND, HAD A LOOK, AND SAID, 'OH WELL, YOU'VE PICKED UP A COUPLE OF FLAK HOLES.' WE DIDN'T KNOW. WE WERE 'GREEN'.

WE PICKED UP OUR FLIGHT ENGINEER HERE. WE WERE ALL IN THIS BIG ROOM, THE CREWS, AND THE FLIGHT ENGINEERS IN THE FRONT. WE SAW THIS OLD BOY WITH A WRINKLED FACE AND I SAID TO MY SKIPPER. 'EH I BET WE GET THAT OLD BUGGER THERE.' WE DID GET THAT OLD BUGGER BUT HE TURNED OUT TO BE THE BEST MATE YOU COULD EVER HAVE – GOOD ENGINEER AND A GOOD PERSON.

WE NEVER WENT FAR WITHOUT EACH OTHER. WE TOOK OUR TRAINING SERIOUSLY. WE MADE UP OUR MINDS THAT WE WERE GOING TO SURVIVE."
DAVE FELLOWES

Above Veteran air gunner Dave Fellowes.

Opposite Air gunner Steve Bethell, third from the left, with his crew, just before take off on the raid to Berlin on 18/19 November 1943.

Gerry Norwood formed up with his crew at the Operational Training Unit at RAF Hixon.

"THEY PUT YOU ALL IN A ROOM, SO MANY PILOTS, SO MANY NAVIGATORS ETC., AND YOU MADE UP YOUR OWN CREW. THEY DIDN'T FORCE YOU TO GO WITH ANYBODY. YOU JUST MIXED IN AND FOUND YOUR OWN CREW.I FOUND A CREW AND WE WERE ON WELLINGTONS DOING CIRCUITS AND BUMPS. I CAUGHT THE RUDDY FLU AGAIN AND WAS SENT TO SICK BAY. THE CREW I HAD PICKED CRASHED ON TRAINING AND WERE ALL KILLED. WHEN I CAME OUT OF HOSPITAL I WENT BACK TO HIXON. I'D GOT NO CREW AND WAS TOLD I WOULD HAVE TO WAIT TO PICK UP ANOTHER CREW. THEN I GOT CALLED UP TO THE ORDERLY ROOM AND THEY SAID, 'WOULD YOU MIND IF WE SENT YOU TO SEIGHFORD?' I SAID, 'WHY?' 'WELL THEY'VE GOT A CREW THERE AND THEY WANT A REAR GUNNER.'" GERRY NORWOOD

Steve Bethell had left No. 4 Air Gunnery School in February 1943 and was sent to RAF Woolfox Lodge and then RAF North Luffenham and No. 29 OTU, where he was crewed up:

"WE WERE ALL PUT IN THIS BIG HANGAR. ABOUT SIX OR SEVEN FULL CREWS BUT NOBODY KNEW EACH OTHER, AT LEAST NOT AS CREWS. I WAS APPROACHED BY A NAVIGATOR. HE WAS A MARRIED MAN AND HE WAS 32, WHICH IS OLD, PROBABLY THE OLDEST MAN IN THE ROOM. HE SAID 'ARE YOU FIXED UP YET?' I SAID 'NO, I'VE ONLY JUST COME IN.' 'WELL', HE SAID, 'WE'VE GOT A REAR GUNNER AND WE WANT A MID-UPPER.' SO HE SAID, 'COME ON I'LL INTRODUCE YOU AND SEE WHAT YOU THINK.' WE

MET AND GOT ON FINE. A VERY NICE CHAP. IT WAS THE SAME WITH THE OTHERS. LATER THE NAVIGATOR TOLD ME HE CHOSE ME BECAUSE I LOOKED LIKE A LITTLE LOST SCHOOLBOY. I SAID, 'I WAS, I WAS.'" STEVE BETHELL

It was during the training phase that the aircrews began to experience the realities of taking to the air in a machine and the risks involved. Ability, teamwork and diligence could mitigate the dangers of defying the natural force of gravity, but one became aware of one's own mortality. One of the many tragic statistics associated with Bomber Command is that around 8,000 airmen lost their lives in training incidents. Pilot George Dunn, having completed a tour of operations, had become an instructor.

THERE ARE TWO WORDS IN OUR VOCABULARY WHICH PROBABLY AFFECT EVERYONE IN THEIR LIFETIME AND THEY ARE FATE AND LUCK, THE FORMER OF WHICH WAS ON MY SIDE ON THE NIGHT OF 20 JANUARY 1944.

I WAS A FLYING INSTRUCTOR ON WELLINGTONS AT AN OTU AND ON THE NIGHT IN QUESTION WAS OFFICER IN CHARGE OF NIGHT FLYING. IT SO HAPPENED THAT ONE OF MY PUPILS WAS HIMSELF A QUALIFIED FLYING INSTRUCTOR HAVING COME FROM FLYING TRAINING COMMAND IN ORDER TO CONVERT ON TO WELLINGTONS AND SUBSEQUENTLY ON TO FOUR ENGINED AIRCRAFT AND THENCE TO A SQUADRON. HE HAD ALREADY GONE SOLO AT NIGHT AND WAS DUE TO CARRY OUT A FURTHER EXERCISE INVOLVING CIRCUITS AND LANDINGS, BUT AS THE VISIBILITY WAS SOMEWHAT HAZY I DECIDED TO FLY WITH HIM ON A CHECK CIRCUIT IN ORDER TO BE SURE THAT NIGHT FLYING WAS POSSIBLE AND SAFE. TO SAVE TIME IN NOT HAVING THE AIRCRAFT RETURN ME TO THE DISPERSAL AREA I ARRANGED FOR TRANSPORT TO PICK ME UP AT THE TAKE OFF POINT WHEN THE PILOT TAXIED ROUND AFTER HIS LANDING.

VISIBILITY WAS QUITE OK, SO I LEFT THE AIRCRAFT AS ARRANGED AND WENT BACK TO FLYING CONTROL. MEANWHILE THE PILOT TOOK OFF AND PROCEEDED TO CARRY OUT HIS CIRCUIT. ON HIS FINAL LEG APPROACHING THE RUNWAY, FOR WHATEVER REASON HE DECIDED TO OVERSHOOT AND GO ROUND AGAIN. THE AIRCRAFT THEN DISAPPEARED FROM THE CIRCUIT, WITHOUT ANY FURTHER COMMUNICATION WITH FLYING CONTROL, AND A WHILE LATER WE RECEIVED A MESSAGE TO SAY THAT AN AIRCRAFT HAD CRASHED AND ALL THE CREW WERE KILLED. THIS TURNED OUT TO BE OUR AIRCRAFT AND A SUBSEQUENT ENQUIRY REVEALED THAT A PROPELLER BLADE HAD BECOME DETACHED CAUSING AN ENGINE FAILURE, WHICH RESULTED IN THE CRASH. WHILE ON THE CHECK CIRCUIT THE ENGINES HAD BEHAVED NORMALLY AND THERE WAS NO INDICATION OF ANY FAULT. HAD THE FAILURE HAPPENED ON THIS CIRCUIT INSTEAD OF A FEW MINUTES LATER I WOULD NOT HAVE WRITTEN THIS ACCOUNT."
GEORGE DUNN

Alan McDonald would go on to serve as a rear gunner with No. 50 Squadron, having been training on Vickers Wellingtons at RAF Market Harborough. On one night-time bombing practice flight Alan recalled: 'We took off and I said to the skipper, "There's a strong smell of petrol in the rear turret." He said, "Well, I don't know where it's coming

Right Veteran air gunner Alan McDonald, who sat in his turret during one flight 'soaked to the skin in petrol'.

from Mac but everything's registering, everything's perfect up front." Then a second time I said, "It's getting stronger skipper." And a third time I said, "It's getting stronger still." And a fourth time I said, "I'm soaked to the skin in petrol skipper." He said, "Oh well we'll make it back to base. We'll cancel the bombing."'

As the aircraft approached the runway at Market Harborough, and having 'passed over some buildings of the 'drome... all of a sudden the aircraft did an about turn. One of the engines cut out. ... We were now going backwards, still going down. We landed in a field and crossed three ditches, the undercarriage standing up to the thumping it got at each ditch, and we ended up in a cornfield. I had turned my turret to beam, and opened the turret door at the back of me. When we touched down and were bouncing along my parachute caught on something, I don't know what. It bellowed out and I got dragged out the rear turret.'

With Alan being, 'soaked to the skin with petrol' he had donned his parachute. 'I thought if we were going to catch fire I didn't want to be anywhere near where there's fire. I'd be the first one out.' Now finding himself some distance from the aircraft Alan picked up his parachute and walked over to the skipper and crew. His captain said: 'What happened to you then Mac? You stink of petrol.'
I said, 'Yeah they'll have me up for pinching aircraft fuel.'

From carrying out training on twin-engined aircraft at the Operational Training Unit – and when the four-engined heavy bombers, the Avro Lancaster, the Short Stirling and the Handley Page Halifax, started to appear at operational squadrons – crews were sent to a Heavy Conversion Unit, to gain a flight engineer and usually a second air gunner, to bring the crew to seven. Navigator Hal Gardner went through the Heavy Conversion Unit at RAF Swinderby, having his first direct experience of the reality of the dangers he was facing:

"WE WERE SLEEPING TWO CREWS TO A HUT AT NIGHT. WE WERE WOKEN UP IN THE MIDDLE OF THE NIGHT WITH RAF POLICE COMING IN AND TAKING AWAY ALL THE OTHER CREW'S EQUIPMENT. SO WE KNEW THEY HAD GONE. APPARENTLY, THEY WERE COMING IN TO LAND, ON THEIR STIRLING, AND ALL FOUR ENGINES SLOWED. THEY WERE ALL KILLED. WE HAD A PROPER FUNERAL FOR THEM WITH UNION JACKS ON THE COFFINS. IT AFFECTED US A LITTLE

BIT BUT WE UNDERSTOOD THESE SORTS OF THINGS. IT DIDN'T MAKE US DESPONDENT, THINKING WE DIDN'T WANT TO GO ON. YOU REALISED WHAT WE WERE DOING COULD BE DANGEROUS. IT DIDN'T STOP US WANTING TO FLY ON." HAL GARDNER

Having formed as a crew, and undergone various navigation, familiarisation and bombing practice flights, it was now time to become part of an operational squadron. The crew may have already ventured beyond United Kingdom air-space, perhaps carrying out a 'Nickel' raid, dropping leaflets over France or Belgium. Now they would journey further over Nazi-occupied hostile territory and to the skies above the Third Reich itself, playing their part in the strategic bombing offensive against Germany.

Air gunner Gerry Norwood arrived at No. 460 (RAAF) Squadron in 1943. Early in his operational career Gerry began to doubt the validity and nature of the bombing offensive he had just become part of:

"WHEN I FIRST STARTED I WAS A BIT DUBIOUS BECAUSE I FELT IT WAS THE WRONG THING TO DO. I WENT DOWN TO THE PUB ONE NIGHT, WE'D DONE ABOUT THREE OR FOUR OPS, AND I GOT TALKING TO ONE OF THE LOCALS. HE WAS TOO OLD FOR THE WAR BUT WAS WORKING ON A FARM. A MARVELLOUS OLD BOY. HE SAID TO ME 'DON'T FEEL GUILTY ABOUT WHAT YOU ARE DOING BECAUSE YOU ARE GIVING TO THE ENEMY WHAT THE ENEMY HAS GIVEN TO US.' FROM THEN ON, I NEVER FELT THAT I WAS DOING WRONG FOR BEING ON BOMBER COMMAND. THE LOCALS – THE RESPECT THEY GAVE YOU WAS MARVELLOUS." GERRY NORWOOD

Left Bomber Command navigator Hal Gardner bottom left.

Right Air gunner Gerry Norwood who would complete a full tour of operations on Avro Lancasters with No. 460 (RAAF) Squadron in 1943/44.

Once at a squadron new crews could ask some of the old lags – 'old' being men in their mid- to late twenties perhaps – what operations were like. They could also read some of the literature describing raids. Maybe their skipper carried out a 'second dickey' operation, flying with an experienced crew and reporting back what he had seen. Pilot George Dunn had actually carried out his second dickey trips while his crew were still at the Heavy Conversion Unit. Shortly after arriving at the HCU at Rufforth, 'and before my feet had hardly touched the ground they said, "You are to go off to No. 10 Squadron at Melbourne and do your two second dickey trips"':

"WHEN I WENT TO BRIEFING IT WAS ALL STRANGE TO ME. I HADN'T EVEN SET FOOT IN A HALIFAX. THE FIRST TRIP, WHERE WOULD IT BE? ESSEN, HOME OF THE KRUPPS WORKS. WE WENT OFF, NO TROUBLE, BUT I COULDN'T BELIEVE WHAT I WAS SEEING WHEN WE GOT OVER THE TARGET. THE FLAK WAS UNBELIEVABLE. THE FOLLOWING NIGHT WAS A TRIP TO KIEL WHICH AGAIN WAS QUITE A HEAVILY DEFENDED TARGET. THEN I WENT BACK TO THE HEAVY CONVERSION UNIT AND MY CREW SAID TO ME, 'WELL WHAT WAS IT LIKE? WERE YOU FRIGHTENED?' I SAID, 'PUT IT THIS WAY. WHEN YOU PACK YOUR GEAR TO GO ON OUR FIRST TRIP REMEMBER TO TAKE A SPARE PAIR OF UNDERPANTS.'" GEORGE DUNN

A RUN OF BAD LUCK

5

June 1943 had been a bad month for No. 460 Squadron, Royal Australian Air Force (RAAF). On the night of the 11th, they lost the crew of Sergeant Robert Christie, shot down by a night fighter over Düsseldorf. The following night they lost another two crews, including the Lancaster of Pilot Officer Lloyd Hadley, DFC. They might have lost another had it not been for the quick actions of the wireless operator in the crew of Flight Sergeant George Cope.

Cope was on the return leg from Bochum when his Lancaster was hit by flak. Moments later, an incendiary bomb came through the side of the fuselage behind the mid-upper turret. They had been struck by a bomb from an aircraft above, an occupational hazard when so many aircraft were over the target at any one time. The wireless operator, Sergeant Douglas Crouch, was quickly out of his seat as the plane caught fire, reaching for an extinguisher. It took several minutes to get the fire under control and a full ten minutes before it was finally extinguished. They landed safely at Binbrook without further incident.

Two nights later and they are operating again. The target is Oberhausen, in the Ruhr, and the squadron has committed twenty-two aircraft to the attack. The crew is halfway through its tour and comprises five Australians, one Canadian and an Englishman. Most have been together since meeting at No. 27 Operational Training Unit (OTU). They have been busy since arriving on the squadron in March; at the time, No. 460 Squadron was based at Breighton in Yorkshire but has since moved.

In April they flew nine operations, including nine-hour long-hauls to Pilsen and Stettin and a successful raid on La Spezia in northern Italy and the inland port of Duisburg. Essen is another favourite; they've been there three times already.

Cope, twenty-three, along with the navigator, Hugh Gordon, are both married, and embarked for the UK in the summer of 1942. Gordon is uncommon in many ways: at thirty-four he is comparatively old for aircrew, and in peacetime he is a qualified barrister and solicitor. Not unusually, perhaps, given his professional qualifications, he is the only member of the crew to hold a commission.

In the small hours of the morning, above the Dutch village of Schinveld, they are stalked by

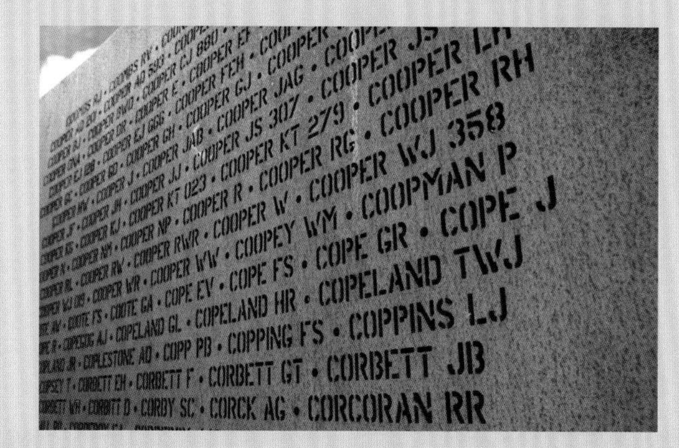

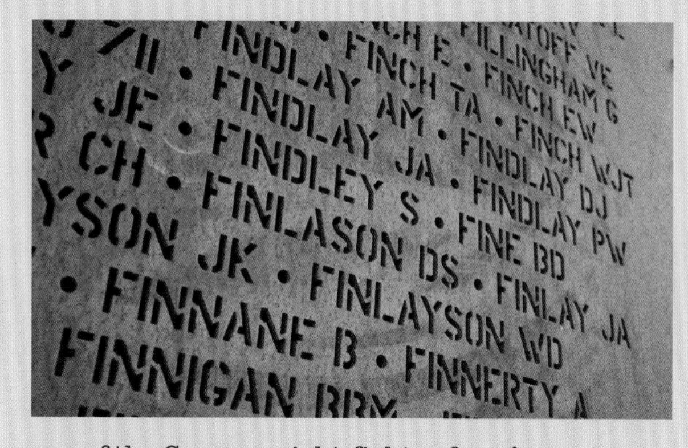

one of the German night-fighter force's most prolific aces. Hauptman Manfred Meurer already has forty-two kills to his name and is looking for his forty-third. He finds it and opens fire.

Soon the Lancaster is falling from the sky, as the crew scramble desperately to leave the stricken bomber. Two, at least, are successful: the flight engineer, Ernie Booth, and the mid-upper gunner 'Bill' Matheson, the Englishman and the Canadian. Both become prisoners of war. The rest of the crew, including 32-year-old Douglas Crouch, the 21-year-old air bomber Douglas Douds and rear gunner Donald Finlason, also 21, are killed.

The squadron's run of bad fortune continues: George Cope and his crew are one of three from Binbrook who fail to return from this evening's raid.

FAILED TO RETURN

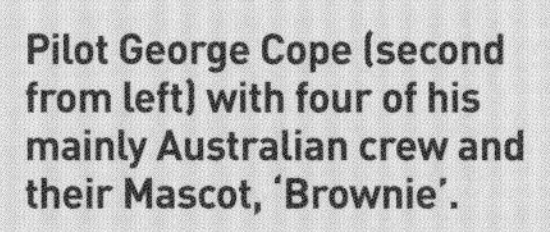

Pilot George Cope (second from left) with four of his mainly Australian crew and their Mascot, 'Brownie'.

BOMBER COMMAND
PART TWO THE AIRCREW STORY

Left Veteran air gunner Steve Bethell at the Spire Memorial unveiling on 2 October 2015 with Lincoln Cathedral in the background. Steve started flying operations with No. 467 (RAAF) Squadron during the 1943 Battle of the Ruhr.

Nothing could prepare aircrew for the actual reality of being part of a bombing raid themselves. What follows are various accounts from the men who did experience that reality and were fortunate to be able to return and share their experiences.

Briefing for a raid followed a fairly standard procedure at the time Steve Bethell was on operations. A map showing the identity of, and the routes to and from, the target, was hidden behind a large curtain when the crews entered the briefing room: 'We had been told in the morning that there were ops on and the time of the briefing. We'd all go in in battle dress. The navigators would go in a little early because they had plotting to do. They were the only ones that knew the target at that time. There would be a little talk, a morale booster about how well we were all doing, and then they would open the curtains. No matter what it was we'd all go, 'Oh God.' Sometimes louder with certain targets than others.'

The crews would be briefed on the route, any areas of defences to avoid, and the expected weather conditions during the raid. 'They would wish us the best of luck and then it was off to have our eggs and bacon. Then to the locker room, get dressed and out to dispersals. Then just wait and hope you'd get off in time.'

In June 1943, Steve, with his skipper Flight Sergeant McLelland and crew, arrived at RAF Bottesford and No. 467 (RAAF) Squadron. Their first operation, to Mülheim, was on 22 June, then Wuppertal on the 24th, Gelsenkirchen on the 25th and Cologne on the 28th. It was as Steve recalls: 'Right in the middle of the Battle of the Ruhr. Quite crowded really. I always thought they'd start us off on an easy one. But they didn't mess around like that.'

"BEING IN THE MID-UPPER TURRET YOU'VE GOT THE BEST SIGHT OF THE LOT. LOOK FORWARD AND YOU CAN SEE EVERYTHING THAT'S COMING. LOOK BACKWARD AND YOU CAN TELL THE REAR GUNNER WHAT'S COMING. ON OUR FIRST OPERATION I COULDN'T BELIEVE IT WHEN WE GOT OVER THE DUTCH COAST. IT STARTED THERE. WE SAW A COUPLE GO DOWN FROM FIGHTERS. THEN IT ONLY SEEMED A FEW MINUTES BEFORE WE WERE THERE, THE RUHR, MULHEIM [SIC], QUITE CLOSE TO ESSEN, AND IT WAS A RING OF FIRE. 'GOD!' I'D NEVER SEEN THAT BEFORE. DESPITE ALL THE TRAINING IT STILL CAME AS QUITE A SHOCK. WE DID GET HIT, RATHER NASTILY, SEVERAL SMALL HOLES AND THE TURRET WAS PUT OUT OF ACTION BECAUSE OF ONE OF THE OIL LEAKS. YOU COULD HEAR THEM GOING THROUGH AND I REMEMBER THINKING THE LAUNDRY IS NOT GOING TO BE

VERY HAPPY ABOUT THIS. WE DID IT, GOT THROUGH IT, BOMBED, APPARENTLY GOT A GOOD PHOTOGRAPH, AND CAME BACK. THE OTHERS IN JUNE WERE SIMILAR BUT NOT SUCH A SURPRISE. BUT ALL RATHER NASTY.

WE HAD .303 MACHINE GUNS AND THEY HAD 20MM OR 30MM CANNON. THAT'S NO CONTEST. THEY HAD LONGER RANGE AND THE MAIN THING WAS TO SEE THEM. I HAD QUITE GOOD NIGHT VISION AND I DID SEE THE ODD FIGHTER BUT NOT EVERY TRIP. WE WERE FORTUNATE ENOUGH TO SEE THEM IN TIME TO GET CORKSCREWING AND GET OUT OF THE WAY. I ONLY OPENED FIRE ON TWO OCCASIONS AND THEY WERE QUICK ONES BECAUSE WE WERE DIVING AND TURNING. BUT WE GOT AWAY AND THAT WAS THE THING. ALWAYS VERY MUCH ON THE DEFENSIVE. IT WASN'T A REAL STAND UP FIGHT REALLY. NOT WITH OUR PEA SHOOTERS AGAINST THEM.

IF THE ENEMY AIRCRAFT IS ON THE PORT SIDE YOU'D GIVE A 'CORKSCREW PORT!'. THAT CONSISTS OF DIVING TO PORT AND THEN DIVING TO STARBOARD, CLIMBING TO STARBOARD AND THEN CLIMBING TO PORT. MAKING A SORT OF DIAMOND. WE'RE TRAINED THAT WHEN WE FIRE, IF THE FIGHTER WAS STILL THERE, WE HAD TO KNOW THE DIFFERENT DEFLECTIONS TO GET TO HIM. I DIDN'T HAVE LONG COMBATS BECAUSE ONCE YOU STARTED THAT CORK-SCREW THERE'S SO MANY OTHER BOMBERS TO PICK FROM HE'D PROBABLY GO FOR SOMEBODY ELSE.

NOW AND THEN THE NAVIGATOR WOULD ASK US TO DO CERTAIN THINGS LIKE CHECK THE WIND GAUGE. BUT BY AND LARGE IT WAS LOOKING OUT. NOT JUST FOR ENEMY AIRCRAFT BUT FOR YOUR OWN. WE SAW SOME NASTY COLLISIONS AND THERE WERE VERY FEW SURVIVORS. WE HAD A LUCKY ESCAPE ON ONE TRIP WHERE WE LOST A WING TIP. THAT WAS OWING TO ANOTHER LANCASTER BEING CHASED. I DIDN'T KNOW WE'D BEEN HIT BUT FELT THIS MOVEMENT. WE DIDN'T PROPERLY REALISE UNTIL WE WERE COMING IN TO LAND AND WE DIDN'T HAVE THE LIGHTS ON THE STARBOARD SIDE. WE WERE FORTUNATE HE HADN'T BEEN ANOTHER YARD OR TWO CLOSER TO US. IN THE DARK, ESPECIALLY IN CLOUD, YOU'D GET THE WASH OF SOME AIRCRAFT BUT YOU COULDN'T SEE IT. YOUR EYES WERE POPPING OUT OF YOUR HEAD. THERE WERE A COUPLE OF HUNDRED OF YOU IN THE WAVE, ALL GOING IN THE SAME DIRECTION AND ALL AT THE SAME HEIGHT. IT WAS A BIT DICEY." STEVE BETHELL

If Steve was flying in thick cloud he recalls that you had no idea if a flash was due to flak or an aircraft exploding. On a clear night, however, it was different: 'I used to consider that I'd see about half of them go. I wasn't always right but if I saw 8 or 9 go I'd say we lost about 20 that night.' Whether or not such sights affected Steve he recollects: 'A selfish mode comes into it. You've got to be honest about that. Hope your luck holds. You sometimes wondered who it was, whether it was somebody you knew. But you didn't have time to ponder on that really. You're looking after your own crew as well as yourself. We always thought of ourselves as a team rather than individuals. We were very very lucky, especially in 1943.'

"I'VE NEVER BEEN A RELIGIOUS MAN. YOU WOULD HEAR SOME PEOPLE SAY, 'I'LL BE ALRIGHT, SOMEBODY UP THERE'S LOOKING AFTER ME.' BUT I NEVER USED TO BUY THAT. PROBABLY BETTER IF YOU COULD. YOU HAD GREAT

POWERS OF RECOVERY AT THAT AGE. AS MUCH AS YOU COULD BE FRIGHTENED OUT OF YOUR LIFE. I DON'T MIND ADMITTING THAT. THERE WOULD BE SOMETHING WRONG WITH YOU IF YOU WEREN'T FRIGHTENED. BUT AS LONG AS YOU KEEP YOUR FRIGHT UNDER CONTROL AND USE THE FRIGHT TO QUICKEN YOUR ACTIONS.

I REALISED THAT I HAD TO CONTROL IT. YOU HAD TO REALISE THAT IT WAS NOW OR NEVER IN SOME SITUATIONS. I'D BE AWARE OF THE RISKS BUT I WAS ABLE TO CONCENTRATE ON WHAT I'D GOT TO DO. I'M GLAD I DID. IT MUST HAVE BEEN AWFUL IF YOU PANIC. YOU MUSTN'T PANIC. YOU AND THE REAR GUNNER – YOU'RE THE EYES. THE ENGINEER KEPT A LOOK OUT AT THE FRONT BUT EVERYTHING HAPPENED TOO QUICKLY IN FRONT. IF ANYTHING IS COMING IT'S FROM THE SIDE AND REAR. THOUGH ONE OF THE WEAKNESSES OF THE LANCASTER WAS NO VISION UNDERNEATH. SO EVERY NOW AND THEN THE PILOT WOULD DROP THE WING AND AS THE MID-UPPER YOU LOOKED DOWN UNDERNEATH. IN THOSE DAYS, IN 1943, MOST OF THE RISK CAME FROM UNDERNEATH. THE NAVIGATOR USED TO CARRY ON LIKE HELL ABOUT IT BECAUSE IT USED TO MESS UP HIS PLOTS." STEVE BETHELL

Steve had his lucky scarf, which went with him on every trip:

"THE BOMB AIMER ONCE FORGOT HIS AND THE PILOT HELD US UP WHILE HE WENT BACK AND GOT IT. INSTEAD OF BEING IN THE FIRST WAVE WE WENT IN THE THIRD WAVE. I DON'T CONSIDER MYSELF SUPERSTITIOUS NORMALLY, BUT WHEN IT CAME TO THE SCARF. WE USED TO GET LEAVE EVERY SIX WEEKS AND I'D PENCIL IN ON MY SCARF WHAT OPERATIONS I HAD DONE IN BETWEEN LEAVE. MY AUNT WOULD THEN EMBROIDER THIS IN.

ONCE YOU LAND, IT'S SO LOVELY TO LAND, AND YOU TAKE YOUR MASK OFF AND YOU CAN HEAR THE SILENCE – PAAAAHHHH – THE AIR – YOU'D HAD THE ENGINE NOISE ALL THE TIME. SEEMED ODD AND WHEN YOU SPOKE YOUR OWN VOICE WAS ECHOING BACK. WITHIN AN HOUR OR TWO YOU'VE HAD YOUR EGGS AND BACON, BEEN DEBRIEFED, GOT BACK TO YOUR BILLET AND WERE HAVING A LAUGH, SAYING IF WE ARE NOT ON FLIGHTS TOMORROW WE'LL GO FOR A JAR." STEVE BETHELL

Jack Pragnell served as an observer with Nos 51, 102 and 298 Squadrons. Jack's twin brother Tom had also trained as an observer with Jack in what was then Rhodesia, but would tragically lose his life on a raid to Berlin on the night of 16/17 December 1943. Jack describes what it was like in the midst of a bombing raid:

"'TARGET AHEAD', THE BLAZING MASS OF SEARCHLIGHTS, THE FLASHING GUNS, THE FLUTTERING IN THE STOMACH? REMEMBER THE RUN IN? STOOGE AROUND LOOKING FOR A CLEARER PATH, WATCHING SOME BRAVE CREW CAUGHT IN THOSE FINGERS OF LIGHT, FLAK GOING UP INTO THAT ENORMOUS CONE, KNOWING THE ALMOST INEVITABILITY OF THAT STREAK OF FLAME HOVERING, HOVERING, PLUNGING VIOLENTLY TO THE GROUND. A QUICK PRAYER THAT THEY GOT OUT. REMEMBER THAT SICKENING LURCH; SLIPSTREAM OR FLAK? OK THANK GOD. MYRIADS OF LITTLE TWINKLING FIRES PEPPERING ON THE GROUND IN SYMMETRICAL PARALLEL LINES, ENORMOUS FLASHES OF EXPLOSIONS. GREEN LIGHTS, RED LIGHTS, BLUE LIGHTS LIKE SOME GARGANTUAN FIREWORK DISPLAY; AND SMOKE,

SMOKE BILLOWING UP IN HUGE CLOUDS, SMALL RINGS OF SMOKE APPEARING ALL AROUND. A QUICK WEAVE AGAIN, OK AGAIN. TARGET AHEAD, BOMB DOORS OPENING, WOULD THEY NEVER OPEN? KITE SAGGING, LISTLESS, LIFELESS. GUNS FLASHING BELOW, TRACER TO STARBOARD. STIRLINGS SILHOUETTED BELOW – BRAVE BOYS THOSE STIRLING FELLOWS. GUNNER SQUALLING TO WEAVE – THAT UNSPOKEN 'MUST BE CALM'. 'LEFT, LEFT' TO THE PILOT. 'FOR GOD'S SAKE SHUT UP' TO THE GUNNER. A BLINDING SILVERING FLASH AS A SEARCHLIGHT CROSSED THE WING. WOULD IT RETURN – SILENT PRAYER – NO IT'S GONE ON. TARGET IN SIGHTS – BUTTON PRESSED, ANOTHER 20 SECONDS STRAIGHT AND LEVEL BEFORE THE PHOTOGRAPH. WOULD THOSE LIGHTS NEVER FLASH, SOMETHING MUST BE WRONG, PILOT WANTED TO KNOW, GUNNER STILL YELLING, CHATTER OF MACHINE GUN, 'COME ON LIGHTS', THERE THEY ARE. BOMB DOORS CLOSED. 'GET THE HELL OUT OF HERE.' OH! BLESSED, BLESSED RELIEF. NOT OUT OF THE WOODS YET BUT OH SO RELIEVED. NOW FOR HOME.

THOUGHTS OF BASE CRAMMED OUR MINDS. REALISATION THAT WE WERE HUNGRY WITH SOME HOURS TO GO YET BUT HOW THE THOUGHTS OF TOMATOES, EGG AND CHIP MEAL MADE THE MOUTH WATER. PILOT RECEIVED A NEW COURSE – ALL BRAKES OFF GOING DOWNHILL MAKING FOR THE COAST. FLAK MILES AWAY TO PORT SHOWED SOME POOR DEVIL IS WAY OFF TRACK GETTING HELL KNOCKED OUT OF HIM. GUNNERS ON THE WATCH OUT ROTATED TURRETS, GRIMLY DETERMINED, WATCHING, PEERING. A QUICK START OF ALARM – NO ONLY DARK CLOUD OR A SPEC ON THE PERSPEX.

'WEAVE SKIPPER, A KITE ON THE PORT NOW. THINK IT IS A LANC BUT NOT SURE.' ALL EYES LOOKING, HOPING. 'YES' MID-UPPER CONFIRMS, 'IT'S A LANC OK.'

FLAK STILL COMING UP ON PORT. A BATTLE ROYAL RAGED AHEAD WITH TRACER FLYING; SUDDENLY A FLASH OF EXPLOSION AND THEN DARKNESS. THE PLANE WEAVED AS THE PILOT MOVED AWAY FROM WHERE THE BATTLE HAD BEEN.

NAVIGATOR YELLING TO KEEP ON TRACK. SKIPPER RESUMES COURSE. NAVIGATOR HAD FORGOTTEN TO SWITCH OFF THE MIC AND CURSED, NO LUCK FROM THE GEE BOX. HAVE TO TRY ASTRO. A CHANCY BUSINESS THAT AND NO FUN SQUEEZING PAST, SHORT OF OXYGEN, DISTURBING ENGINEER BUSY COMPUTING AND WITH THE SKIPPER YELLING TO PUT OUT THAT LIGHT, BUT A QUICK SHOT WAS TAKEN. NOT A BAD SHOT THAT CONFIRMS THE TRACK. COAST AHEAD. ON THE WAY OUT IT LOOKED MUCH MORE PLEASANT THAN ON THE WAY IN. NOT MUCH HOPE OF A

Left Twins Jack (left) and Thomas Pragnell. Thomas would lose his life on the raid to Berlin on the night of 16/17 December 1943.

PINPOINT; THE COASTLINE IS TOO STRAIGHT. NAVIGATOR GIVES CHANGE OF DIRECTION ON DEAD RECKONING AND TIMING TO COAST, PLUS THAT ASTRO SHOT AND THEN GIVES A WHOOP, 'GEE BOX WORKING.' SOMETHING MORE CONCRETE THAN FAITH TO GET YOU HOME BY.

WATCH OUT FOR FLAK SHIPS. ALL WAS OVER BAR THE SHOUTING. WE ALL KEPT FINGERS CROSSED FOR CLEAR WEATHER AT HOME, NO DIVERSIONS, NO BALLOONS, FOR PETROL LASTING OUT AND A QUICK PERMISSION TO LAND WITH NO STOOGING AROUND FOR HOURS AVOIDING OTHER CLOTS CIRCUITING. HEIGH HO FOR THAT MEAL. GUNNER BURST OUT INTO 'SPRINGTIME IN THE ROCKIES'. COFFEE WENT ROUND AND FRIENDLY BLACK WATER SHOWED BELOW.

IT WAS GOOD RETURNING TO THAT FRIENDLY STRIP OF TARMAC AND TAXYING ROUND UNDER ORDERS FROM THE GROUND CREW. IT WAS GOOD CLIMBING INTO THAT SMELLY BUS, PICKING UP CREW OF 'G' GEORGE AND MAKING OUR WAY TO THE OPS ROOM. GOOD IN THAT COLD HALF LIGHT TO DRINK WARM COFFEE, TO SEE THE BOARD OF RETURNED CRAFT, WATCHING FOR ANY OVERDUE. WE SAT THERE TALKING, EXHILARATED, BUT SLIGHTLY SUBDUED, WIDE AWAKE BUT BARELY REALIZING THAT ALL WE HAD SEEN WAS REAL. IT HAD BEEN STRANGELY ALOOF UP THERE WITH ONLY THE ROAR OF THE ENGINES AS AN ACCOMPANIMENT. NO HEAT FROM THOSE FIRES OR SOUND FROM THOSE BOMBS HAD REACHED US, YET DEATH HAD WAITED VERY, VERY CLOSELY AT HAND AND HAD BEEN THWARTED. SOME HAD SUCCUMBED. FRIENDS WOULD BE ABSENT AND PANGS OF SORROW WOULD BE FELT. WE LIVED, HOWEVER. AND LIFE WAS GOOD. I HAD A DATE THAT NIGHT, OPS PERMITTING, AND I WOULD DRINK A FEW TOASTS TO THOSE EMPTY PLACES, AND AS WITH GREATER MEN – 'SO TO BED'". JACK PRAGNELL

Air gunner Gerry Norwood would complete a full tour of duty during the late 1943 and early 1944 Battle of Berlin period, and recalled an incident returning from a raid to Magdeburg in the early hours of 22 January 1944:

"WE WERE ON THE OUTER CIRCLE AT BINBROOK AND THE SKIPPER CALLED UP TO SAY WE WERE IN THE FUNNEL AND WE WERE TOLD TO PREPARE TO LAND. THEN THE FOUR ENGINES STOPPED. THAT WAS IT. THE SKIPPER SAID 'HANG ON CHAPS, WE'RE GOING IN'. WE COULDN'T HAVE BAILED OUT. WE WERE UNDER 5,000 FEET COMING IN TO THE FUNNEL. THE SKIPPER MUST HAVE SEEN THE GROUND COMING UP AND AT THE LAST MINUTE HE MUST HAVE PULLED THE STICK BACK AND LIFTED THE NOSE UP, AND HIT THE DECK TAIL FIRST. IF WE HAD GONE IN HEAD FIRST I DOUBT WE WOULD HAVE SURVIVED. WE DIDN'T SET ON FIRE BECAUSE THERE WAS NO PETROL. WE HIT TAIL FIRST AND THE REAR SPAR BROKE. THE TURRET TURNED OVER AND THE GUNS DUG IN TO THE GROUND. THEY CAME THROUGH AND KNOCKED ME STRAIGHT THROUGH THE DOORS – BURST THE DOORS. I WAS JUST LAYING LOOKING UP AT THE SKY. I'D GOT BLOOD COMING OUT OF MY MOUTH AND THEY THOUGHT MY RIBS HAD GONE INTO MY LUNGS. IF THAT HAD HAPPENED THEY COULDN'T HAVE DONE MUCH. THEY RUSHED ME TO LOUTH AND THE INFIRMARY. THEY SENT A TELEGRAM HOME TO MY MOTHER BECAUSE THEY THOUGHT I HAD HAD IT. FORTUNATELY IT WASN'T THAT BAD. I WAS VERY LUCKY. THE BOMB AIMER WAS BADLY INJURED WITH A COUPLE OF BROKEN RIBS." GERRY NORWOOD

Following the incident Gerry was asked if he wanted either to go back and pick up a new crew or stay on the squadron and be a spare. Along with his friend, Ronnie Mansfield, he said he would stay on the squadron, 'because of the local people', and it was made clear to Gerry that he would take the place of someone in a 'sprog crew' on their first operation.

"THE FIRST OP WAS THE WORST PART BECAUSE YOU DIDN'T KNOW IF ANYONE WAS GOING TO PANIC. YOU ONLY WANTED ONE TO PANIC AND THE REST WOULD FALL IN. ONCE YOU START PANICKING THAT'S IT. YOU MAY AS WELL SAY THAT'S THE END.

THE FIRST SPROG CREW I FLEW WITH, I SAID TO THE SKIPPER, 'DON'T THINK OF ME AS BRASH OR BRAVE OR TRYING TO FORCE YOU TO DO ANYTHING BUT NOBODY SHOULD TALK UNLESS THEY'VE GOT TO. BUT YOU MUST REMEMBER THAT EVERY TWENTY MINUTES OR SO YOU SHOULD CALL UP THE CREW THAT YOU CAN'T SEE. YOU CAN SEE THE ENGINEER SITTING NEXT TO YOU. YOU CAN SEE THE WIRELESS OPERATOR, BUT YOU CANNOT SEE THE REAR GUNNER, OR THE MID-UPPER OR THE NAVIGATOR, SO YOU CALL UP AND MAKE SURE THEY ARE ALRIGHT AND THERE'S NO PANIC. THE ONLY OTHER THING I CAN TELL YOU IS WHEN YOU APPROACH THE TARGET YOU DON'T HAVE TO FLY STRAIGHT AND LEVEL. YOU CAN DIVE BECAUSE THE QUICKER YOU GET IN TO THE TARGET AND OUT THE LESS CHANCE YOU'VE GOT OF BEING HIT. YOU COULD BE HIT WITH SOME OF YOUR OWN BOMBS COMING DOWN FROM AIRCRAFT ABOVE YOU. PUT YOUR NOSE DOWN AND BELT LIKE HELL THROUGH THE TARGET. EVERY CREW I FLEW WITH – THEY ALL FINISHED A TOUR.

Left Veteran Gerry Norwood at the Spire Memorial. Gerry flew a tour of operations as an air gunner with No. 460 (RAAF) Squadron. 'If you saw a Lancaster being attacked by a fighter you could see the rear gunner and the tracer bullets coming out and whipping all over the sky. You never seemed to think that it might be you.'

WE WERE OVER BERLIN AND THE MID-UPPER GUNNER STARTED FIRING. WHEN I SWUNG ROUND TO SEE WHERE, I SAID, 'STOP FIRING MID-UPPER. IT'S ANOTHER LANC.' FORTUNATELY IT WAS FAR ENOUGH AWAY. THE .303 BULLETS ONLY USED TO GO A FEW YARDS AND THEY'D START GOING ALL OVER THE SKY. THE .303S WERE NOT POWERFUL ENOUGH. UNLESS THEY WERE ABSOLUTELY ON TOP OF YOU YOUR OWN GUNS WERE USELESS. THEY DID TRY .5 BROWNINGS BUT THEY WERE SO POWERFUL THEY BROKE THE MOUNTINGS. SO THEY HAD TO GO BACK TO .303S.

MY JOB WAS TO MAKE SURE YOU KEPT AWAKE AND WATCHFUL – THAT YOU KEPT SEARCHING UP, DOWN, PORT, STARBOARD. MAKE SURE THAT NO AIRCRAFT WAS NEAR YOU. ON TOP OF THAT WATCH FOR PREDICTED ANTI-AIRCRAFT FIRE BECAUSE THE GERMANS HAD MARVELLOUS FLAK. YOU'D SEE, SO MANY YARDS AWAY, A SHELL BURST. THEN A FEW MOMENTS LATER ANOTHER ONE WOULD BURST BUT MUCH NEARER TO YOU. SO IF YOU SAW IT CREEPING UP ON YOU YOU HAD TO CALL THE PILOT AND SAY STANDBY TO DIVE EITHER PORT OR STARBOARD. IT DIDN'T MATTER WHICH WAY YOU TURNED AS LONG AS YOU TURNED OFF, BECAUSE THE NEXT ONE WOULD BURST ON TOP OF YOU.

ON DISPERSAL, WAITING TO TAKE OFF, YOU WERE STANDING OUTSIDE THE AIRCRAFT AND THE DOCTOR WOULD COME ROUND IN HIS CAR AND GIVE YOU YOUR WAKEY WAKEY PILLS AND ENERGY PILLS. AS SOON AS HE TURNED HIS BACK A LOT OF GUNNERS WOULD SPIT THEM OUT. A SILLY THING TO DO BECAUSE IF THE REAR GUNNER FELL ASLEEP THAT WAS FATAL. TWO OR THREE TIMES AIRCRAFT TRIED TO TAKE OFF AND EXPLODED ON THE RUNWAY – DID A GROUND LOOP AND BLEW A HOLE IN THE RUNWAY. OPS WERE SCRUBBED AND THEN AT TWO OR THREE IN THE MORNING YOU'D STILL BE IN THE MESS BECAUSE YOU COULDN'T GO TO SLEEP BECAUSE YOU'D TAKEN YOUR WAKEY WAKEY PILLS.

ONCE YOU SPOTTED AN ENEMY AIRCRAFT ALL YOU HAD TO DO WAS MAKE SURE THAT HE WAS FLYING STRAIGHT AND LEVEL. IF HE WAS YOU'D SAY TO THE MID-UPPER TO WATCH THE AIRCRAFT. TELL HIM WHERE IT WAS AND AS SOON AS HE HAD GOT IT TO KEEP HIS EYE ON IT. IF HE STARTED TO DIP HIS WINGS HE WAS TO LET ME KNOW BECAUSE HE COULD BE THE DECOY AND WHILE YOU WERE WATCHING THAT ONE ANOTHER ONE WOULD COME UP FROM UNDERNEATH OR FROM ABOVE. IF YOU GOT YOUR MID-UPPER TO WATCH THAT ONE YOU SEARCHED THE SKY TO MAKE SURE THERE WAS NO ONE ELSE THERE.

YOU WOULD SEE OTHERS CAUGHT IN THE MASTER SEARCHLIGHT, THE AIRCRAFT GLINTING IN IT. THE NEXT MINUTE ABOUT 30 SEARCHLIGHTS WOULD COME UP AND ALL CROSS WHERE HE WAS, AND THEN THE ANTI-AIRCRAFT WOULD START COMING UP. IF YOU WERE CAUGHT IN THE MASTER BLUE SEARCHLIGHT THEN DON'T TRY TO WEAVE. THE ONLY WAY TO GET OUT WAS GO STRAIGHT DOWN NOSE FIRST, TURN ROUND AND GO BACK, AND THEN TURN ROUND AND GO FORWARD. YOU'D SEE THE ANTI-AIRCRAFT AND THE NEXT MOMENT WHOOMPH, HE'D GONE POOR BLOKE.

IF YOU SAW A LANCASTER BEING ATTACKED BY A FIGHTER YOU COULD SEE THE REAR GUNNER AND THE TRACER BULLETS COMING OUT AND WHIPPING ALL OVER THE SKY. YOU NEVER SEEMED TO THINK THAT IT MIGHT BE YOU. IF YOU HAD YOU WOULD PROBABLY HAVE GONE LMF [LACK OF MORAL FIBRE]. IT DIDN'T SEEM TO REGISTER THAT YOU COULD BE NEXT. I DON'T THINK I EVER SAW

ANYBODY REALLY AFRAID. YOU USED TO TALK BUT NEVER SAY THAT YOU COULD BE NEXT OR IT MIGHT BE YOUR TURN TONIGHT. THEY ALL SEEMED TO HAVE MASCOTS. ONE BLOKE HAD A SCARF THAT HE WOULDN'T FLY WITHOUT, OR GLOVES. YOU BECAME VERY SUPERSTITIOUS. YOU HAD TO HAVE YOUR MASCOT AND ONCE YOU HAD THAT YOU WEREN'T AFRAID." GERRY NORWOOD

When speaking to veterans of Bomber Command, they will often remark on how their survival can be attributed in large part to pure chance. Gerry Norwood recalls the tragic loss of his friend Ronnie Mansfield, along with all his crew, as they prepared to raid Villeneuve on 9/10 April 1944:

"RONNIE WAS MY FIRST WIRELESS OPERATOR AND HE STAYED ON AS A SPARE. WHEN I CAME BACK FROM THE OPERATION THE CREW BUS CAME OUT TO PICK US UP. NEXT MINUTE A CAR CAME OUT AND THE WAAF DRIVER SAID TO ME 'COME HERE.' I ASKED WHAT SHE WANTED, AND SHE SAID, 'I'VE GOT TO PICK YOU UP.' 'WHAT FOR?' I ASKED. SHE REPLIED, 'DUSTY MILLER', THE SIGNALS OFFICER, 'ASKED ME TO PICK YOU UP AND MAKE SURE YOU GOT BACK TO THE INTERROGATION WITH NO PROBLEM.'

THE OP WE HAD BEEN ON WAS A VERY EASY OP IN FRANCE. WHEN WE TOOK OFF, WE TOOK OFF FIRST AND WERE CIRCLING THE AERODROME TO GET HEIGHT AND SET COURSE. WE HEARD RONNIE'S CAPTAIN CALL UP THE CONTROL TOWER ON THE RT, 'I CANNOT BRING MY PORT WING UP.' HE PANICKED AND THEY ALL STARTED TO PANIC. YOU COULD HEAR THE SHOUTING. THE CONTROL TOWER WAS TRYING TO GIVE HIM INSTRUCTIONS ON WHAT TO DO TO BRING HIS PORT WING UP. THEY WENT AROUND IN EVER DECREASING CIRCLES AND BOOM, BLEW UP.

DUSTY MILLER HAD TO DO SO MANY HOURS TO GET HIS FLYING PAY AND ON THIS PARTICULAR OP HE TRIED TO TALK RONNIE INTO CHANGING. HE'D SAID, 'RON, STAND DOWN. I'LL GO IN YOUR PLACE BECAUSE I WANT TO GET MY FLYING HOURS IN.' RONNIE SAID, 'NO, I'M GOING.' IF HE HAD CHANGED DUSTY WOULD HAVE GONE. THAT'S WHY HE SENT A CAR OUT. HE WAS FLABBERGASTED. HE CAME TO ME AND SAID, 'I DON'T KNOW WHAT TO SAY. RON WOULDN'T CHANGE. IF I'D HAVE CHANGED THAT WAS IT.'" GERRY NORWOOD

This and other losses had a marked effect on Gerry's attitude towards friendship: 'I was very reluctant. I thought what's the point of trying to make friends. They'd suddenly get killed and it might affect you even more. Never deep friends. Didn't want it. You might say right I'm not going to fly any more. So rather than that don't make friends.'

Ken Johnson flew as an air gunner on Lancasters with Nos 9 and 61 Squadrons and recalls an incident on 25 July 1944 during a raid to Saint-Cyr:

"APPROACHING THE TARGET A LANC ABOVE OPENED HIS BOMB DOORS TO REVEAL TWO ROWS OF RUSTY BOMBS; I WARNED OUR SKIPPER BUT HE SAID WE WERE HEMMED IN, ADDING HE WILL SEE US AND NOT DROP THEM (WISHFUL THINKING). AT THIS TIME I HAD TO BE QUIET AS THE BOMB-AIMER GUIDED THE PILOT TO THE TARGET. AS OUR BOMB-AIMER CALLED 'BOMBS GONE' SO DID THOSE IN THE PLANE

Right Lancaster air gunner Ken Johnson.

Below Veteran air gunner Ken Johnson looks on to the name of his fellow gunner Carson John Foy, who lost his life in a tragic incident on 25 July 1944.

ABOVE. QUICKLY THEY FLIPPED FROM HORIZONTAL TO VERTICAL. MOST WENT BETWEEN THE TAIL AND WING, ONE HOWEVER STRUCK THE STARBOARD RUDDER, BREAKING THIS OFF THEN CONTINUING ON ITS WAY. OUR PLANE BUCKED AROUND AS THE SKIPPER FOUGHT TO REGAIN CONTROL. THE INTERCOM WENT DEAD SO I WAS UNAWARE OF WHAT WAS HAPPENING. WE SEEMED TO BE STEADY SO I SAT TIGHT. NEVER IN MY TWENTY YEARS HAD I FELT SO LONELY BUT I HAD TO STAY ALERT IN CASE OF FIGHTERS. AS I TURNED FORWARD I FOUND A SECOND BOMB HAD BROKEN FOUR FEET OFF THE WING, WHICH DANGLED, LIKE A BIRD WITH A BROKEN WING. THIS FELL OFF SOON AFTER.

A LITTLE WHILE LATER I REALISED I HAD NOT SEEN THE REAR TURRET MOVING. AS HE MOVED SIDE TO SIDE HIS GUNS WOULD BE SEEN AND THIS WAS NOT HAPPENING. THE MID-UPPER WAS BACKLESS, SO BY POSITIONING FORWARD AND BENDING DOWN TO LOOK BETWEEN MY LEGS I COULD SEE DOWN THE FUSELAGE. TO MY DISMAY THERE WAS A GAPING HOLE. I THEN REALISED WE HAD LOST JACK. MY STOMACH TURNED OVER. I KNEW WE HAD LOST HIM FOREVER IN THE MOST HORRIBLE WAY POSSIBLE. I BEGAN GLANCING TOWARD THE COCKPIT WHERE OTHER CREW MEMBERS WERE, HOPING TO ATTRACT ATTENTION. EVENTUALLY THE ENGINEER PUT HIS HEAD UP INTO THE ASTRODOME CHECKING DAMAGE AND I MANAGED TO SIGNAL HIM. TO GET TO ME HE HAD TO OPEN A DOOR WHICH BLOCKED THE REST OF THE CREW FROM THE TWO GUNNERS. MINUTES LATER HE TUGGED AT MY TROUSER LEG AND BECKONED ME TO BEND SO HE COULD YELL IN MY EAR. HE HAD MY PARACHUTE FROM THE REAR END, 'THE SKIPPER WANTS YOU TO WEAR THIS IN CASE WE HAVE TO LEAVE IN A HURRY', ADDING WE WILL WARN YOU

IF WE HAVE TO JUMP. IT MUST HAVE BEEN ANOTHER HOUR BEFORE THE INTERCOM SPLUTTERED INTO LIFE, THE ENGINEER HAD GOT IT WORKING. THIS WAS A BLESSING FOR ME. AT LAST I FELT IN TOUCH WITH THE OTHERS.

NOW NEARING THE FRENCH COAST, THE SKIPPER ASKED FOR A ROUTE TO AN EMERGENCY CRASH DROME NEAR RAMSGATE WHICH HAD AN OVERSIZED RUNWAY TO HELP DAMAGED AIRCRAFT TO LAND SAFELY. WE CROSSED THE CHANNEL, WHEN THE SKIPPER CALLED TO THE NAVIGATOR, 'I HAVE BROUGHT IT THIS FAR. GIVE ME A HEADING FOR HOME.' THE RADIO OPERATOR SIGNALLED BASE ABOUT OUR DAMAGE. FATE AGAIN INTERFERED. AS WE GOT WITHIN SIGHT OF LINCOLN CATHEDRAL WE WERE REFUSED PERMISSION TO LAND. AN AIRCRAFT AHEAD HAD CRASHED, BLOCKING ONE RUNWAY AND WE WERE DIVERTED TO AN AIRFIELD NEAR NEWARK. THE SKIPPER, AS AN ANTI-CLIMAX, MADE A PERFECT LANDING. HE WAS TOLD TO TAXI TO THE CONNING TOWER AND TURN AROUND BEFORE STOPPING ENGINES. THE SILENCE WAS UNBELIEVABLE. NO ONE SPOKE AS WE VACATED THE PLANE. WE DID NOT TAKE A LADDER WHEN FLYING AS THIS, BEING MADE OF METAL, WOULD HAVE INTERFERED WITH THE COMPASS. I WOULD NORMALLY HAVE JUMPED THE THREE FEET TO THE GROUND BUT ON THIS OCCASION MY LEGS WERE SHAKY. HAD I JUMPED I FELT I WOULD FALL ON MY FACE SO I WAITED FOR THE GROUND STAFF TO LEND A HAND. STILL NO WORDS WERE EXCHANGED, THEY SEEMED TO FEEL THE TENSION. THE SKIPPER WAS LAST TO LEAVE. HE FLEW IN SHIRT-SLEEVES, UNLIKE THE TWO GUNNERS WEARING A LOT OF EXTRA CLOTHING TO TRY TO KEEP WARM, AND HIS SHIRT WAS SOAKED IN SWEAT. HE HAD FOUGHT ALL THE WAY HOME WITH LITTLE MORE THAN HALF THE CONTROLS. HE WAS AWARDED THE DISTINGUISHED FLYING CROSS FOR HIS EFFORTS. STILL NO WORDS WERE EXCHANGED UNTIL THE STATION CO ARRIVED IN HIS CAR. HE WAS VERY CONCERNED ABOUT OUR PLIGHT AND REMARKED THAT ONE LITTLE THING HE WOULD DO WAS ENTERTAIN US TO A SLAP-UP MEAL IN THE OFFICERS' MESS. HARDLY HAD HE UTTERED THE WORDS WHEN A DISPATCH RIDER ARRIVED, CLOSELY FOLLOWED BY A CREW BUS. WE HAD TO RETURN AT ONCE FOR DE-BRIEFING, AND WE DID NOT GET THE MEAL. NEXT MORNING WE WERE DRIVEN BACK TO OUR PLANE FOR A PHOTO SHOOT. AS I WAS THE SMALLEST I HAD TO CREEP ALONG TO THE DAMAGED TAIL POINTING TO THE REAR GUNNER'S CHUTE STILL IN ITS STOWAGE. HOW IT STAYED IN THAT POSITION WAS A MYSTERY. RETURNING TO BASE FOR LUNCH WE FOUND OUR NAMES ON BATTLE ORDERS FOR THAT NIGHT, WORKING ON THE OLD ADAGE, IF YOU FALL GET UP AT ONCE." KEN JOHNSON

Ken's rear gunner colleague, Canadian Carson Foy, had not survived the incident and rests in Fontenay-le-Fleury Communal Cemetery.

Luck certainly had no respect for rank or experience. Steve Bethell had completed his first tour with 467 (RAAF) Squadron on the raid to Berlin on 16/17 December 1943. For many aircrew the magic number was thirty operations, although those serving as Pathfinders were expected to fly more, and in the run up to D-Day some operations over France were classed as one-third of an op owing to the shorter time over hostile territory. As one Canadian gunner, Jack Fitzgerald, wrote to his mother: 'Each trip consists of only 1/3 of a op, which is silly because if you go

Opposite Veteran navigator Hal Gardner.

ATKINSON JB • ATKINSON JC • ATKINSON JM • ATKINSON JW
ATKINSON MR • ATKINSON RAP • ATKINSON RG • ATKINSON RH
ATKINSON TD • ATTER WL • ATTERBY E • ATTFIELD RA
ATTWELL KD • ATTWOOD CAV • ATTWOOD JH • ATWOOD BE
AUDLEY DS • AUGUST DC • AULD C • AULD J • AULT RE
AUMELL AD • AUSTEN FJ • AUSTEN WG • AUSTIN AS • AUSTIN D
AUSTIN GJ • AUSTIN H • AUSTIN HW • AUSTIN J 933
AUSTIN JH • AUSTIN JP • AUSTIN KH 833 • AUSTIN KH 425
AUSTIN LH • AUSTIN M • AUSTIN WE • AUSTIN WEJ • AUSTIN WF
AUSTIN WH 987 • AUSTIN WS • AUTY E • AVER FA
AVERY RAC • AVEY GR • AVIET PB • AVON JLG • AWAD PP
AXFORD NF • AXUP GA • AYERS RB • AYERST FD • AYLARD AC
AYRES AWW • AYRES GA • AYRES GW • AYRES KA • AYRES L
AYRES LR • AYRES RF • AYRES RK • AYRES VH • BABER AT
BABIACKI S • BABIARZ J • BABINGTON-BROWNE KD • BABRAJ Z
BACKHOUSE FL • BACKHOUSE L • BACKLER AF • BACKLER H
BACKS WH • BACKSHALL EO • BACKWAY HG • BACON DA
BACON L • BACON LW • BACON PF • BACZKIEWICZ E
BADCOCK R • BADDELEY DH • BADDELEY RA • BADGE H
BADGER VJ • BADLEY JW • BADOWSKI Z • BAGG RPM
BAGLEY AE • BAGLEY D • BAGLEY JW • BAGLEY RW
BAGNALL RS • BAGSHAW C • BAGSHAW G • BAIGENT PH
BAILES RJ • BAILEY A • BAILEY AJM • BAILEY AR • BAILEY C
BAILEY D 025 • BAILEY D 449 • BAILEY DJ • BAILEY E 826
BAILEY EA • BAILEY EH • BAILEY GC • BAILEY GI • BAILEY GJ
BAILEY HJ • BAILEY HR • BAILEY J 915 • BAILEY J 774
BAILEY P • BAILEY PAG • BAILEY PF • BAILEY R • BAILEY RAG
BAILEY RW • BAILEY SAJ • BAILEY SL • BAILEY T
BAILEY VL • BAILEY W • BAILEY WJ • BAILLIE T
BAIN A • BAIN CGS • BAIN FGL • BAIN JS • BAIN R • BAINBRIDGE GB
BAINES C • BAIRD A • BAIRD AGG • BAIRD DM • BAIRD DR
BAIRD R • BAIRD RJ • BAIRD T • BAIRD WA • BAIRD WOD
BAIRSTOW EA • BAIRSTOW JL • BAK J • BAKE AG
BAKER A • BAKER ACH • BAKER AD • BAKER AF • BAKER AS
004
BAKER RB
BAKER T
BAKER WJ 345
BALA E • BALA ES
BALDWIN AH
BALDWIN J
BALDWIN RW
BALFOUR DC
BALKWILL HJ
BALL EF • BALL F
BALL L
BALL RW 260
BALLANTYNE AG
BALLARD PS
BALLEINE BGP
BALMER J
BALUCKI J
BAMFORD J
BANFIELD AJ
BANKOWSKI PR
BANKS IW
BANKS RC
BANNISTER DP
BARACZ EJ
BARBER BR
BARBER FG
BARBER JL
BARBER RT
BARCLAY A
BARCLAY WB
BARFOOT EE
BARKER A 771
BARKER DM
BARKER FF
BARKER JW

over there and get killed you don't get only 1/3 killed.' Jack lost his life in August 1944. Steve Bethell would actually fly twenty-five operations on his first tour. 'For the Aussies, with a small population, to lose a crew was a bigger loss to them than to a country with a 50 million population. So at 25 operations they packed us in instead of the 30.' Steve was sent to take up instruction duties at No. 1660 Heavy Conversion Unit, mainly lecturing. Then in June the opportunity for a second tour arose: 'A friend of mine, Rowland Middleton, another gunner, was due to go back. He had been off for well over a year. He said to me, "Steve, we're quite good friends. What about coming back? I'm due to go back with Pop Atkins." You can tell by the name he was old. He didn't want to go back. He was married and I suppose he must have been all of 31. That's old. The last thing you want to do is go back on ops with someone who is reluctant.'

Steve's response was clear: 'No. I've just packed in and I'm just getting used to this.' He was told that a Wing Commander Walker, who had earned a DFM earlier in the war, was looking for a second crew and that Pop would be very pleased. "He'd buy you a beer every night if you took his place." I said, "No, not yet. I would if I'd done a few more months but not now." Anyway that night we went out and had a few jars and by about 9 o'clock I said "Yeh, of course I'll come." I saw old Pop and said I'd take his place. He wanted to kiss me.'

Steve began his second tour with No. 49 Squadron at RAF Fiskerton in July 1944, carrying out some daylight raids as part of the V-weapons counteroffensive. 'I was

Left Veteran navigator Henry Wagner. Henry's Halifax was shot down on the night of 17/18 December 1944, with Henry the only survivor from the crew of seven.

optimistic at this time. It's so much easier of a day for a gunner. You look round and can see for miles. Mind you there was some good flak and the frightening thing was the collisions. I saw a few.' In September 1944 Steve was sent to No. 83 Squadron, part of 5 Group's No. 54 Base at RAF Coningsby to carry out Master Bomber duties under the auspices of the legendary pilot Guy Gibson of Dam Buster fame. On 9 September Steve and crew flew a Master Bomber familiarisation with Gibson on board (ten days later Gibson would lose his life). However, Steve, shortly after this flight, became ill. 'It was the only time I was grounded, with a sinus problem, and a chap named Dunn took my place.' On 11/12 September the crew went missing on a raid to Darmstadt. Wing Commander Walker, his navigator, and the two gunners, Rowland Middleton and Steve's replacement Cyril Dunn, lost their lives. The remainder of the crew became POWs. Steve recalls: 'It was the only one I missed. I had been in hospital for a couple of days. I remember going back to the billet and my logbook had been taken and put in the archives. It went with the other logbooks, which they take away when someone's missing. They had got my logbook and it had gone up with all the deads. I had to write to get it back.'

The experiences we have read about are a mere snapshot of the experiences of the aircrews, seeing aircraft blow up, tending wounded crew mates, witnessing the demise of colleagues. And they would be experiencing this extreme one moment, staring death in the face, and within a few hours they would be back at base having a meal, and soon, perhaps, be in a pub chatting with a female companion, laughing with their friends. Then go through it all again, and again, and again. If they were lucky to survive. The official history of the Strategic Bomber Offensive stated that sixty per cent of bomber crews became casualties, 55,573 airmen did not survive the war, 9,838 became prisoners and 8,403 were wounded. Recent research carried out by the International Bomber Command Centre team is revising this figure, upwards, in terms of aircrew lives lost and including ground crew and Women's Auxiliary Air Force personnel. The scale of loss is higher than at first thought.

Post-war recognition of what the personnel of Bomber Command endured was certainly not forthcoming. The controversy over the bombing policy arrested any acknowledgement of the aircrews' airmanship, risk-taking and sacrifice, and manifested itself in the absence of a specific campaign medal after the war and Winston Churchill ignoring them in his victory speech, despite earlier in the war claiming that 'The fighters are our salvation ... but the bombers alone provide the means of victory.' Towards the end of the war plans were devised for a wide-ranging examination of the conduct of the bomber war and its effectiveness. The Air Ministry were certainly keen. But the extent of the plan was questioned by Churchill, who called it a 'sterile task' and eventually a greatly reduced survey was carried out. Nothing like that conducted by the Americans – the comprehensive United States Strategic Bombing Survey. Chief of the Air Staff Sir Charles Portal would write that, because of this, Government opinion in the UK in regard to the air offensive – and this author also suggests public opinion – would be 'based largely on propaganda, personal recollection or on the results of investigations of other nations'. So there became a void when it came to the objective analysis of the RAF bomber offensive. Subjective agendas therefore filled this space, which perhaps should always begin with the concept of intentionality when it comes to the bomber war and the effect on those who took part.

Having served with the Royal Air Force's Bomber Command, these airmen, from numerous nations allied to the fight for freedom, returned to their ordinary lives – careers, family, occasionally getting together with their former colleagues. And when they did there was no celebration of the act of war and killing, but pride in defeat of Nazism, and the lifelong friendships they had made. They will say they were only doing a job. They will also commemorate those lost, but they know the job had to be done. They will never forget those times, as navigator Jack Pragnell recalls:

SO MANY MEMORIES JOSTLED ONE ANOTHER. SOME MORE VIVID, SOME LESS MARKED, BUT ALL OVERSHADOWED BY THE SEARING MEMORIES OF LIFE ON BOMBER COMMAND. I SUPPOSE THEY WERE THE MOST VIVID BECAUSE THEY WERE MEMORIES OF THE EASILY IMPRESSIONED MIND OF A YOUTH, FACING HAZARDS, FACING DEATH, NOT WITH EQUANIMITY BUT OFTEN WITH FEAR GNAWING THE SOUL, OFTEN WITH AN INEXPRESSIBLE ELATION. PERHAPS IT WAS THE MEDIUM OF FIRE AND DISASTER THAT MADE THOSE IMAGES SO INDELIBLY IMPRINTED.

I REMEMBER THOSE DAYS, MORE PARTICULARLY NIGHTS, WHEN WE, THE CREW, MET LIFE WITH ELAN THAT SURPRISED THE FOE AND AMAZED THE FRIEND. THE CREW – THAT LOOSE-KNIT SET OF INDIVIDUALS ON THE GROUND BUT IN THE AIR AN ENTITY WITH ONE COMMON OBJECT, TO SET OFF, TO BOMB AND TO RETURN TO BASE TO ENTER IN THE LOGBOOKS THAT RATHER TERSE LACONIC STATEMENT, 'OPS. BASE ... BASE – D.C.O (DUTY CARRIED OUT).'

THE CREW WAS TYPICAL OF HUNDREDS, HALF VETERAN, HALF LIKE MYSELF RAW; HALF COMMISSIONED, HALF N.C.O. ALL WERE HUMAN WITH HUMAN FAILINGS AND WEAKNESSES. NO SUPERMEN MADE UP THESE CREWS, JUST ORDINARY FELLOWS FROM ALL WALKS OF LIFE, THE CLERK, THE MECHANIC, THE REGULAR, THE STUDENT, THE RICH MAN, THE POOR MAN, YES AND THIEF.

FROM WHAT HIDDEN SPRINGS DID THOSE FELLOWS DRAW IN ORDER TO FACE DEATH AGAIN AND AGAIN IN ORDER TO RETURN TO A HELL ON EARTH; EACH MOMENT MAY BE THE LAST?; TO RETURN IN THE COLD SOBER LIGHT OF REALISATION AFTER THE FIRST FLUSH OF YOUTHFUL FIRE AND IMPETUOSITY HAD BEEN DAMPED. WAS IT ALL SHEER COURAGE?

COURAGE THERE WAS IN GREAT AMOUNT. I AM CERTAIN THAT EVERYONE POSSESSES SOME DEGREE OF COURAGE UNKNOWN UNTIL NEEDED. IT WAS NOT ALL SHEER 'GUTS'. EACH CREW, AFTER THE FIRST OPS, HAD AN IDEA OF THEIR OWN INVINCIBILITY AND IN EACH OF US THE FEAR OF LETTING HIMSELF DOWN IN FRONT OF OUR FRIENDS WAS IMMENSE." JACK PRAGNELL

There is certainly a change in recent times towards raising the profile of the bomber offensive and remembering and

Below Bomber Command veteran Jack Smith who served as a wireless operator during the latter stages of the war. Jack was involved in operations to repatriate Allied POWs at the end of the war, 'This made a very pleasant change and the former POWs were naturally in good spirits.'

commemorating those who took part. The control of the telling of their story has shifted, with voice to those who were there, who knew someone, who were related to a participant. Across the country numerous memorials have been erected at former bomber stations or at aircraft crash sites, and in 2012 the Bomber Command Memorial in The Green Park, London, was unveiled by Her Majesty Queen Elizabeth II. And now the remarkable International Bomber Command Centre, overlooking Lincoln and the cathedral, takes the recording and telling of the human story of Bomber Command to another, unprecedented, level. Bomber Command veteran George Dunn DFC L d'H attended the unveiling of the International Bomber Command Centre Spire in 2015.

"IT IS DIFFICULT TO FIND THE RIGHT WORDS TO EXPRESS HOW I FELT WHEN I ATTENDED THE UNVEILING. THE ELEGANT SPIRE ON TOP OF THE HILL LOOKING ACROSS THE VALLEY TO THE DISTANT CATHEDRAL WAS A VERY EMOTIONAL MOMENT, ESPECIALLY SURROUNDED BY THE METAL PANELS SHOWING THE NAMES OF THE AIRCREW OF BOMBER COMMAND WHO DIED IN THE SECOND WORLD WAR. SINCE THE UNVEILING I HAVE PAID A VISIT TO THE CHADWICK CENTRE, WHICH DISPLAYS ITEMS AND PHOTOGRAPHS FROM THE CENTRE'S EXTENSIVE ARCHIVES, AND TELLS THE STORY OF THOSE WHO TOOK UP THE STRUGGLE DURING THE WAR. ALL THIS IS NOW AVAILABLE FOR FUTURE GENERATIONS TO SEE AND APPRECIATE THE SACRIFICE MADE BY ALL THOSE WHO SERVED WITH BOMBER COMMAND." GEORGE DUNN

Veteran Gerry Norwood also attended the unveiling.

"I WOULD LIKE TO THANK THE PEOPLE OF THE IBCC FOR THE WORK THEY HAVE DONE. THEY HAVE ERECTED A MEMORIAL THAT, IN YEARS TO COME, YOUNG PEOPLE CAN GO TO AND READ THE NAMES OF THE YOUNG MEN AND WOMEN THAT VOLUNTEERED AND PAID THE ULTIMATE PRICE. THEY ARE ON SHOW. THIS COUNTRY HAD DONE AS MUCH AS IT COULD TO PUSH BOMBER COMMAND UNDER THE CARPET. BOMBER COMMAND WAS A BAD WORD, OR IT SEEMED TO BE FOR POLITICIANS. BUT WHEN YOU LOOK BACK ON IT BOMBER COMMAND DIDN'T START THE WAR AND THE ENEMY STARTED THE BOMBING. IT'S A VERY GOOD THING THAT SOMEBODY HAS NOW COME FORWARD TO MAKE SURE THAT THE YOUNG MEN AND WOMEN WILL NEVER BE FORGOTTEN." GERRY NORWOOD

A LONG WAY FROM HOME

6

Flight Lieutenant Arndt Walther Reif is a small man with an enormous personality. He is unusual in that he left his home in Baker, Oregon, to join the Royal Canadian Air Force (RCAF) and is now serving in Bomber Command. He is also different in that his father is from Dresden. His crew includes an Englishman, George Owen, the wireless operator; a Scotsman, John Paterson, the flight engineer; and four Canadians – Peter Uzelman and Ken Austin, the air bomber and navigator respectively, and the two air gunners, Bob Pearce and John MacLellan. They adore him. They have been together since the start of 1944, having come together at an OTU and a Heavy Conversion Unit (HCU), and flew their first operation with No. 101 Squadron shortly after D-Day.

Earmarked as an exceptional crew, they are asked to join the Pathfinder Force and arrive at No. 582 Squadron at Little Staughton in August. The squadron has only been operating since April and has already established a proud reputation. By December they are an experienced PFF crew, and are briefed to take part in an attack on the marshalling yards at Cologne. The operation is briefed and postponed several times before they are finally cleared to go.

They've been to Cologne before, in October, for

Left Walt Reif (third from right) an American whose father came from Dresden.

their twentieth operation, and nearly didn't make it back. They had landed back at base with two engines dead, 66 holes in the port wing, 153 hits in the body, and the Lancaster's instruments and controls extensively damaged. Now they are asked to go in daylight, with the promise of thick cloud cover to protect them. It will be their thirtieth trip. That will leave fifteen still to do, for Pathfinders fly a 'double tour' equating to forty-five operations.

It's called a 'heavy Oboe' raid. A small force of Lancasters will mark the target using the Oboe blind bombing device usually reserved for Mosquitoes. Indeed, the Mosquitoes are there in reserve, should something go wrong. It requires the leader to fly straight and level during the bombing run, to ensure total accuracy. But it goes wrong from the start. There are no clouds. The sky is clear blue, and the flak gunners cannot believe their luck as the heavy force lumbers towards them. Then a squadron of Luftwaffe fighters appears, led by a leading 'ace', and chaos ensues.

At the front of the formation is Squadron Leader Bob Palmer. He has flown more than 100 trips. He sticks resolutely to his task. When he bombs, the rest of the formation will bomb in salvo. But already some of the Lancasters have started to scatter. Walt Reif is in P-Peter, not his usual aircraft. He stays with his leader, and drops his bombs in salvo. Then there is a terrific 'thump' as cannon shells explode in the tight confines of P-Peter's cockpit. Walt is badly hurt, and his voice can only be heard faintly on the intercom, ordering his crew to bail out. The navigator cannot find his 'chute. His desperate voice is the last to be heard.

Out of the stricken bomber only two parachutes emerge, that of the two gunners, John MacLellan and Bob Pearce. Both are Canadian. Bob Pearce, the rear gunner, only survives by dint of wearing one of the seat-type 'chutes worn by pilots. He turns his turret so that the doors are outside the aircraft and drops out. Of the rest of the crew there is no news. They will never return.

THE INTERNATIONAL BOMBER COMMAND CENTRE
PART THREE THE DIGITAL ARCHIVE

From the start of the project, the team were clear that they would not be developing a museum with artefacts in glass cases. The venture was about building a source of information that was accessible to people across the world and which could be used to educate generations to come about those who served and died in Bomber Command. It was to be about the people who flew and maintained the aircraft, those who looked after the crews and those who were affected by their actions.

The digital archive is run in conjunction with the University of Lincoln, with additional funding from the Heritage Lottery Fund. The university has extensive skills and facilities that make the archive possible, including specialist historical researchers, computer systems and links across the world, which will ensure the archive is a useful resource rather than simply a collection of names and data.

The scope of the digital archive project is at the same time immense and a race against time. By collecting and curating the memories and personal histories of those involved in Bomber Command, it forms one of the key pillars of the project.

Originally the archive was intended to be a repository for the oral histories and stories of the veterans that the team could trace, and a few logbooks and other documents. The size of the task, however, rapidly became evident and the archive soon contained over 600 interviews and over 100,000 scanned documents, photographs and diaries etc.

A particular challenge (as with the losses database) has been the struggle to get stories from people other than aircrew – many of whom have thought that no one wants to hear their story, as is often reflected in the general lack of recognition of their efforts. While the majority of the aircrew are happy to be interviewed, many of the ground crew feel their role was not as important, despite the fact that there were around ten ground crew for each aircrew member. The archive needs to reflect this and there is a massive task still to make sure the balance is right.

The archive team has a database of around 2,000 people they want to interview – a number with only sketchy information such as a name and location. As those in Bomber Command came from across the globe, the recording of the oral histories is also an international undertaking, with volunteers working as far afield as Australia and New Zealand to collect the stories of those who served.

One of the most important features of the project is the aspect of reconciliation, so the team have also started to interview those who were on the other end of actions, such as Luftwaffe pilots, anti-aircraft gun crews and European civilians. Through the international connections of members of the team, they now have over fifty interviews that have been recorded with Italian nationals who experienced the bombing, all of whom have a story to tell.

It is crucial for the team that those recording the oral histories show the story of the people involved and their lives, not just the time they served in Bomber Command. Interviewers are trained to talk about their life stories – their families, their schooling and their reasons for signing up as well as their operational activities. Just as importantly they are interested in what happened after the war – what the veterans went on to do and how they coped with the memories they had forged during active service. In fact, the volunteers often discover that veterans are as proud of their pre- and post-war achievements as they are of their actions during their service.

Just as vital to the archive are the 'paper' records – the documented source material such as the logbooks and other operational information that comes into the project in a number of ways. Ever mindful of the value of the artefacts to the families who own them, there is a strict protocol for the collection, digitisation and return of items to ensure that they remain protected at all times.

To some extent the scanning of documents and recording of the interviews is the most straightforward part of the archive project. The 'real' work starts when the

Previous page IBCC volunteers at the Memorial Spire. Their work has been an essential part of bringing the project to life.

Left Archaeology during the groundwork at the IBCC site.

Opposite Operation Manna Floral Tribute Display, April 2015

information has to be transcribed and catalogued. This is a massive task. For every minute of digital history, for example, the transcription takes around ten minutes by a team of dedicated volunteers.

All images are scanned in at the highest resolution possible – the 'preservation masters' – which become the fall-back originals should they be required. These need to be 'future proofed' so they will be compatible with evolving technology and so are stored in a variety of methods. There are then a series of smaller, watermarked, lower-resolution images, which are the ones available via the website.

Attaching the correct particulars to the objects in the archive is very important as without this the archive would lack the facility to be usefully searched. So, for instance, the information for a flight logbook would include who the person is, what squadrons they served on and between which dates.

The other important element of engaging volunteers is ensuring that they use the same vocabulary when describing images, so a 'control vocabulary' has been designed to make certain that descriptions are as uniform as they can be. This control vocabulary also certifies that the data is accurate, as volunteers are trained only to input information they are clear on. For example, if they know a plane is a Lancaster, but not whether it's a Mk I, Mk II or Mk III, they are advised to just say 'Lancaster', as at least this is correct information for the database.

The volunteers are based across the world – the great advantage of being a digital archive. Training is now undertaken online via videos on the IBCC website, which again ensures consistency whether the interviews are taking place in Malaysia or Milton Keynes! Many volunteers have family connections to Bomber Command, others are keen enthusiasts and countless are learning as they go along. Ages range from early twenties to sixties and seventies, but they are all united by a desire to help the project achieve its objectives.

With letters and diaries, volunteers are sent .pdf files of the items, which they then transcribe for the project, again making them searchable on words, and retrievable as a document or as the original pdf.

The aim at the moment is to gather as many artefacts as possible, particularly with the oral interviews; the transcriptions are a longer-term part of the plan.

Unlike the National Archives, which record the views and thoughts of the national leaders, the IBBC scheme is chronicling the stories of the people on the front line – the

Left Inside the exhibition centre, bringing to life the stories collected via the digital archive project.

Opposite Poppies adorn one of the Walls of Remembrance panels.

flight sergeants, ground crew and others whose accounts would otherwise be lost: anecdotes and memories from families, and others who have owned them for years, many of which have not seen the light of day since the war. It's as much about social and cultural history as it is a military history project, and with no real idea about how people in the future will use the data, it's about collecting and cataloguing as much information as possible.

Bringing the Data 'Alive'

The information in the archive has been used to develop the centre's exhibition. Video stories have been recorded using actors the same age as when the stories were taking place to bring alive the memories and make them 'real'.

With one of the aims being the focus on education, the display has to take account of a wide range of audiences, from schoolchildren to veterans. Some will be visiting to learn, others to confirm what they already know, but careful crafting and selection of the accounts will ensure that they all gain something from their trip.

The exhibition has a handful of object cases and one of the challenges of being a digital archive is selecting the objects to display. Tony Worth was quite clear from the outset of the project that he did not want the Centre to be about 'bent and twisted pieces of metal' – it's very much about the people and their stories and the archive will ensure that this is the case.

The international aspect of the undertaking is also

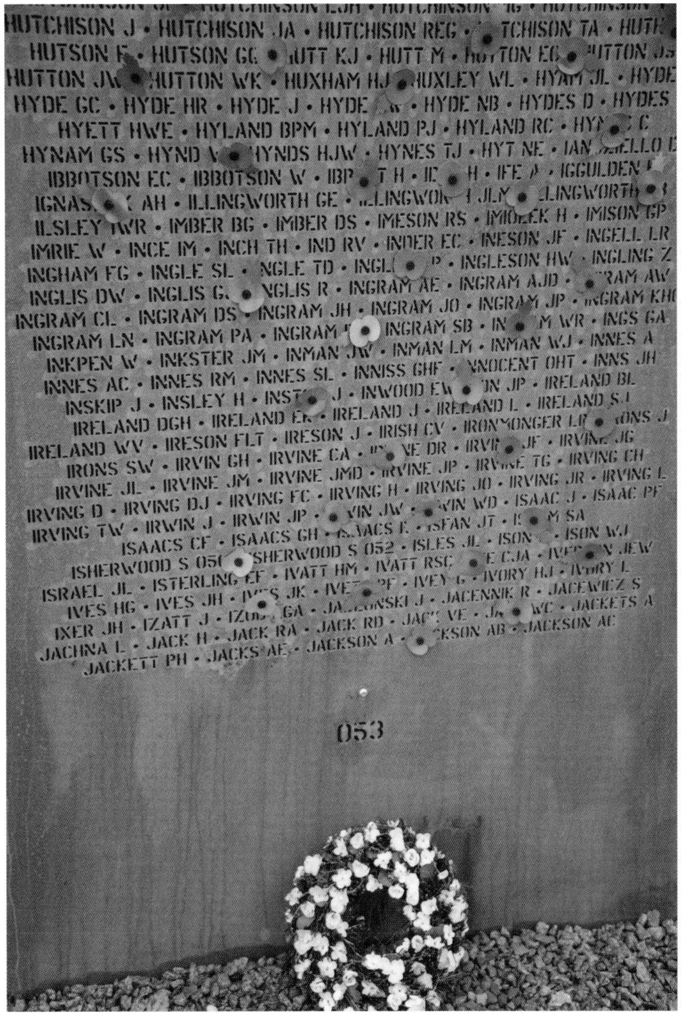

reflected in contributions to the archive, which come from history groups and universities in Italy, the Netherlands, Denmark and Canada, to name but a few. Inroads are also being made in Germany, in line with the 'reconciliation' objectives as contact is made with museums and archives. In terms of challenges while building the archive, there have been many. The sheer scale of the project, which has grown over time, has been a particular issue, but the overarching challenge has been the urgency of the task, in particular with the collections of the oral histories, the source of which by their very nature is a diminishing pool. Funding bodies, however, have been acutely aware of this and have been understanding in helping the enterprise continue its valuable work while decisions are made.

Although the archive holds thousands of files, there are many more items to be added – some are discovered by accidental conversations, others by detailed research; but the project team are determined to collect and store the memories of veterans' service to Bomber Command.

While the digital archive and losses database are two of the centrepieces of the venture, the physical aspects are just as important and the most immediately visible elements of the team's endeavours.

DECISION OVER PFORZHEIM

The Grundy crew. The navigator, Roy Sowter (second left) insisted his skipper went first.

The raid on the town of Pforzheim on the night of 23 February 1945 has been devastatingly successful. More than eighty per cent of the town's built-up area has been destroyed. Captain Ted Swales of the South African Air Force (SAAF), the Master Bomber, call-sign 'black tie', has done his job well, although the raid will be the last for this heroic South African. Twenty-seven Lancasters from No. 550 Squadron out of North Killingholme are included in the Main Force, and while the flak is comparatively light, the night fighters are especially active.

One Lancaster, NF998 D-Dog, is in trouble, albeit from a danger closer to home. Less than thirty seconds after bombing, a shower of some fifteen incendiaries from a 'friendly' aircraft above has struck them a near-fatal blow. The mid-upper gunner is hurt, and one of the bombs is embedded behind the port outer engine and burning fiercely. Rudder, elevator and trim controls are damaged, and the hydraulic lines severed. The Canadian pilot, Flying Officer Robert Harris, on only his fifth operation, faces a long and dangerous flight home on three engines. He will survive, only to be killed the following month over Dessau.

Elsewhere in the stream, the air bomber in veteran Lancaster LM273 has also successfully found the target and the navigator is plotting their route home. Suddenly, this Lancaster too is in trouble, and falling, thirty or so miles to the south-west of Stuttgart. It has been shot out of the sky with a sudden violence that has caught the crew completely unawares. Even experienced men in a tight-knit crew can sometimes fall foul of a night fighter and, later, Oberfähnrich Helmut Bunje will be celebrating his third victory of the night.

In the Lancaster, the skipper, Flying Officer D.H. Grundy, tries at first to extinguish the flames that are licking at his port wing, but to no avail. He knows it is hopeless and gives the order to bail out, continuing to call his crew on the intercom. The rear gunner, Sergeant Ernest MacKenzie, says he is on his way but there is no response from his fellow gunner, Sergeant Ernest Jarvis, in the mid-upper turret. The wireless operator, Flight Sergeant Leslie Figg, reports that the turret has been hit, cannon shells and machine-gun fire from their assailant smashing through the middle fuselage. He thinks Ernie is done for.

The pilot stays at his controls for as long as possible and leaves his seat at the very last moment to make for the escape hatch and plummet into the night sky. Fortunately, he's wearing a seat-type parachute, rather than the more cumbersome chest-type version. His fellow Canadian and navigator, Pilot Officer Roy Sowter, is only moments behind, having insisted his skipper goes first. A decision had to be made and there is no time to argue. The crew have been together for more than twenty operations and are two-thirds through their tour. This will be their last.

Grundy pulls at his D-ring and there is a sharp and reassuring crack as his parachute opens above him. He sees his burning bomber continue to fall in a shallow right-hand dive. He hits the ground shortly after. At least four other members of the crew have made it out of the stricken bomber. The flight engineer, Sergeant E.W. King, is in considerable pain, having seriously broken his ankle, but at least he is alive. It is an injury from which he will never fully recover. The air bomber, Roy McLauchlan, is also safe, but has a nasty bullet wound in his right lung and is gasping for breath. Les Figg is unharmed but quickly captured.

Of the 21-year-old navigator, there is no immediate sign. He will be found dead, his parachute seemingly having failed to deploy, perhaps through lack of height. The two air gunners will also later be reported killed in action.

RAF BOMBER COMMAND GROUND PERSONNEL
THREE CHEERS FOR THE MEN AND WOMEN ON THE GROUND

'Erks' (as ground personnel were colloquially known) and WAAFs, the women of the Women's Auxiliary Air Force, were an essential part of Bomber Command. They worked long hours in often challenging conditions to keep aircraft serviceable and to look after the well-being of the flyers. They accounted for more than eighty per cent of the population on an operational station, but were sometimes disparagingly referred to as penguins, wingless wonders, or the chair force. To some extent, this perception of ground personnel has continued to this day. Often overlooked, surviving ground personnel are modest and still seem reluctant to tell their stories. To date, the IBCC Digital Archive has recorded almost 1,000 oral history interviews with veterans, but less than a tenth of them are with ground personnel or WAAFs. When they are persuaded to tell their life stories to the IBCC interviewers, several themes emerge. Whatever their trade, ground personnel and WAAFs talk of their experiences during training, their working and living conditions when posted to operational stations, what they got up to on leave, and the relationships that they formed during the war.

Many men chose the RAF because they wanted to fly; being part of a modern force that relied on cutting-edge technology was appealing. Some who fell short of the stringent tests to become aircrew consoled themselves with being close to the aircraft. Women joined for similar reasons, but after conscription was introduced in December 1941, there was a rush to volunteer for the WAAF. Winifred Barker volunteered just before her twentieth birthday: 'I decided I would like to join the Air Force and if I waited any longer I would be called up to do munitions or Land Army, both of which I didn't really take a fancy to.'

Like thousands of other women, Winifred thought she would rather serve in the WAAF than be conscripted to work in a factory. Others have admitted they were influenced by the colour of the uniform. Once in the Air Force, the iron beds and mattresses made up of three 'biscuits' were some of the first things new recruits had to contend with, and feature in many of the IBCC Digital Archive's memoirs and oral histories: 'Every morning we had to fold our blankets, [stack] our biscuits, put the blankets round the biscuits and do like everybody else had to do, towel and irons for inspection, hiding our laundry out of the sight of the NCO.'

It was an important moment when they were first issued with their uniform and kit and sent their civilian clothes home. It signified that they were now part of the RAF: 'You went to the counter, there was a chap there, he just said, "Tunic, size off"... he never measured you. ... Then you went and changed, then you parcelled your own civvy clothes up and sent them back home, you were in then. ... Two tunics, two pairs of trousers, pants, vests, socks, and they were all coarse in those days, and the boots ... most people had never worn boots ... you had to get used to the boots.'

The routine medical inspections and inoculations came as a surprise for recruits. Margaret Young described them: 'We had injections which weren't very pleasant ... you had a needle put into your arm ... you left the needle in and moved on to the next doctor ... another injection and another dose ... and also we had what we called FFIs – Free From Infection – to see if you had anything in your hair or any skin disease. ... You had to strip to the waist ... and be examined by doctors to see if you were alright.'

The men went through the same procedures: 'Line up and pants down. ... The bigger men seemed to pass out more than the others, a lot of men passed out at the thought of inoculations.'

Depending on when they joined, basic training consisted of attending lectures and up to eight weeks of physical instruction and drill, or 'square bashing' as it was called. New recruits might have an idea of which trade they would like, but it was by no means guaranteed. Charles Bland was lucky. He was accepted for his preferred trade 'You'd put down on your exam, on your joining what you'd like to be. ... I wanted to be an engine fitter. Some wanted to be riggers, armourers and instrument makers.'

After basic training, the trade training course they were posted to depended on their results in aptitude tests and which trades were most urgently required by the RAF. Instruction was not specific to any command. Possibly because she had learned Morse code as a Girl Guide, Margaret Young trained for a further six months as a wireless operator at Blackpool and Compton Bassett in Wiltshire. She remembers that she made friends, 'but not lifelong friends because you were all going in your separate directions'. Whatever their trade, personnel could be sent anywhere. Throughout their Service careers, friendships were broken up and new ones created at the next location. Being posted could be daunting. Separated from those she knew, Edith Kup found false bravado was necessary: 'In rather a quake I arrived in the mess, always an ordeal at a new station, for the boys ... were waiting to see what presented itself – I used to stand outside the anteroom door, count to ten and then stride in, head held high and looking straight ahead – a quick glance to locate any WAAF and go over and join her. Once that was over there were no further qualms. One was quickly accepted as part of the team.'

Despatched to No. 11 Operational Training Unit at RAF Oakley, Winifred Barker soon settled into Service life as a stenographer and made friends with the other WAAFs in K hut. She recalls: 'Pat was born in New Zealand, Phyllis from Canvey Island, Ada from Walthamstow, Rita from Guernsey and Peggy from Oxford. We all went to Peggy's

Previous page **No. 50 and 61 Squadron Armourers at RAF Skellingthorpe.** (The Cluett collection)

Opposite

1. Aero engine fitter Bill Marshall with his Hawker Hind, RAF Waddington, 1938. (The O'Flanagan collection)

2. Loading 1,000 lb bombs onto a No. 420 Squadron RCAF Halifax probably at RAF Tholthorpe, 1944. (The Jones collection)

3. RAF Middleton St George's photographic section, Bill Maxwell, Chris Carruthers, Arthur Carr, Reg Cavalier and Bill Welesby wait to develop target photographs. (The Cavalier collection)

4. Cookhouse staff, RAF Skellingthorpe, 1942. (The Cluett collection)

5. A WAAF at RAF Waltham/Grimsby. This picture is from an album rescued from a skip and given to the IBCC. (The Otter collection)

6. Refuelling No. 467 Squadron's Lancaster S-Sugar at RAF Waddington. (The Hughes collection)

1

2

3

4

5

6

wedding to her Polish pilot Stefan.'

People from diverse backgrounds were accepted in the RAF; their trade and their ability to work efficiently became an important part of their identity. WAAFs were pleased to be issued with trousers as they denoted a technical trade, and ground crew were proud to be 'scruffy erks'. Consequently, new personnel wanted to lose quickly their 'sproggy' appearance and to look like they'd 'got some in'. With time, the blue of their uniforms faded, their footwear was broken in and, after hours of polishing, uniform buttons and badges began to lose their sharply defined edges. One WAAF tried to get worn 'tapes' to sew on her uniform when she was promoted so she would look like she had more seniority.

On a typical station in 1944, over 1,500 personnel of all trades worked together to keep two squadrons of aircraft and around 200 aircrew operational. Posted to RAF Elsham Wolds, Ted Mawdsley explained that

> **Ground crew would outnumber [the aircrew] by six, seven times ... you had the engine fitters, engine mechanics, airframe fitters, the armourers who did all the bombing up ... all who actually worked on the aircraft. But in addition to that, you had people in the cookhouse, you had clerical workers who worked in admin. You had drivers ... you had motor mechanics and you had people who looked after the petrol depot and also where the bombs were kept. And batmen who looked after the officers and so on and so on.**

Left **A typical scene in a Nissen hut as groundcrew gather round a stove. Temperatures in the semi-cylindrical corrugated steel structures, widely used on RAF stations, varied considerably in line with the seasons.** (The Simpson collection)

Each Bomber Command station was like a large village, but the living conditions were basic, especially on newly built, dispersed sites: 'We had appalling bloody conditions to live in, Nissen huts with a stove pipe in the middle, and if your bed happened to be by the stove pipe, well God help you, because everybody sat on your bed to get as near to the fire as they could.'

Personnel had to walk or cycle miles between their living quarters, the ablutions (the washing facilities), the mess, and the sites where they worked. Margaret Bailey remembered: 'The WAAF sites were in a wood ... they were Nissen huts and you'd hear the rats running over the top at night ... you'd have to get out to go to the toilet in the dark ... but that was life, you just took it as a matter of course.'

Most personnel accepted the conditions with the typical wartime stoicism. The 'Waafery' could be a twenty-minute walk from the technical site, but for those who worked on the aircraft 'on the flights' the distances were greater. Travelling tens of miles a day by bicycle was the norm on some dispersed sites: 'We used to peddle out there day and night … my aircraft was billeted on the far side, it'd be two miles … and of course there were no lights, the 'drome was completely in darkness … but nobody ever moaned about it.'

Work on the flights was demanding. In the early years of the war, a large ground crew worked on each aircraft, but by 1942 the expansion of Bomber Command meant that at RAF Waddington, mechanic Eric Howell worked on his No. 44 Squadron Lancaster with just two fitters and a rigger. For a short period of time he had to carry out the daily inspections on the eight engines of two aircraft by himself. It was hot in summer and freezing in winter. At RAF Bourn, Cambridgeshire, with No. 97 Squadron, Kenneth Locke-Brown explained: 'It was very cold … particularly if you were sat up on the wing of a Lancaster filling it with petrol, which [took] three hours … it [was] bloody cold. It [was] cold to sit on it, you [were] elevated, and the weather in winter was very severe indeed … all we had was a leather jerkin and a naval type roll neck sweater … other than that, just overalls over our ordinary uniform.'

Meal times were frequently missed because of their work. If they were lucky, sandwiches and hot drinks were sometimes sent out to them, and when they eventually got to the mess, ground crew often had to make do with left-overs. However, the pride that ground crew felt is evident. In the culture of the wartime RAF, they owned the bombers; the aircrew merely 'borrowed' them. If an aircrew managed to complete their tour they would be on an operational station for a matter of months, but ground personnel could serve on the same station for years. Consequently, they had a clear picture of the losses and attrition of aircraft and aircrew. Ted Mawdsley developed a rapport with the crews of his aircraft and felt it keenly if they failed to return. Eric Howell lost ten Lancasters over two years; like all ground crew, when his aircraft failed to return he worried he had missed something in his inspections that had led to each aircraft's loss. For Dennis Brett, a flight mechanic at RAF Carnaby emergency landing ground, crash-landings were a regular occurrence: 'Some aircraft burst into flames when landing. Others were already on fire as they approached. The sight of a red gun turret is one that I cannot forget. Even our medical officer was seen to turn pale sometimes.'

WAAFs also saw the attrition rate at first hand. Some dated aircrew who all too often did not return from operations. Many of their trades were also vital to the safety of aircrew. As an instrument repairer, Gladys Gildersleve knew that if she made a mistake the aircrew would be 'up the creek', but she also remembers a WAAF who took her trade seriously enough to test one of the parachutes she had packed. Ground personnel roles were not without risk. Around 1,000 ground personnel and over 70 WAAFs died in service with Bomber Command. All are remembered on the IBCC memorial.

For all ground personnel, the cramped conditions in the living quarters facilitated the spread of illness and infection. For those on the flights, working long hours and under pressure to keep all aircraft serviceable, fatigue could lead to mistakes and injuries. Loading bombs with a hand winch was exhausting. One armourer remembers it took something around twenty turns of the handle to lift a bomb an inch, and it was 'heartbreaking' if the load had to be changed. Bombing up Stirlings was particularly hard because of their height. Although power-operated winches were introduced, the weight of the bomb loads also increased, and the systems driving the winches were often temperamental. No. 15 Squadron at RAF Mildenhall in Suffolk had twenty-two aircraft and three bombing-up teams; the first team to finish their seven aircraft had to start loading the last. For armourers, there was always the possibility of a catastrophic accident. The bomb load of a No. 514 Squadron Lancaster at RAF Waterbeach exploded on 29 December 1944, destroying the aircraft and damaging seven others on their nearby dispersals. It is thought that a defective bomb caused the sympathetic detonation of the aircraft's entire bomb load. Of those working on the aircraft or in the vicinity, nine were killed and four were wounded.

Other trades were less dangerous but equally necessary. The work of kitchen staff was gruelling and WAAF cooks earned less than half the average weekly wage. Some resented the fact that many of their tasks were used as 'jankers' (punishment for minor misdemeanours). Their day began at 05.00 hours, heating 40 gallons (180 litres) of porridge in giant cauldrons and frying hundreds of rashers of bacon in huge square pans for breakfast. All ground personnel justified their part in the war effort by their proximity to the aircraft and interaction with aircrew; kitchen staff rationalised that they cooked the aircrew's bacon and eggs. At RAF Pocklington in Yorkshire, Edith Kup worked a twenty-four-hour rota system in the watch office. She was often on duty from seven at night until nine the next morning. When a 'Queen Bee' administration officer submitted a new rota, the watch-keepers carried on as before. Kup claims she preferred to work a long night shift as it enabled her to 'carry through the operation from start to finish'. Working these shifts, WAAFs were able to maintain their contact with the flyers.

Although it was one of the lowest paid trades, driving was another rewarding but emotionally charged role on an operational station. Drivers like Margaret Bailey were among the last people to see flyers as they were driven to the dispersals before an operation: 'We had little Bedford vans with a canvas roof at the back ... we used to take them out in that ... and we used to be there with them until they took off. ... We'd sleep in a little hut ... then they'd call us and we'd have to get up and go and bring them in and then we'd go back to our billets.'

Methods of travel to and from leave, between postings and on nights out were also memorable events for IBCC interviewees. Joyce Bell was a WAAF stationed at Bawtry Hall near Doncaster: 'I remember standing with three other WAAFs ... waiting for a bus when a man driving a horsebox pulled up and asked if we would like a lift. ... We gladly accepted his offer ... [and] jumped in beside the horse. The driver dropped us off right outside the cinema where people were queuing. The expression on their faces when four WAAFs jumped out was hilarious.'

The population of a typical bomber station consisted of young men and women, many of whom were single and had recently left home for the first time. Dancing, drinking, the cinema and romance were frequent pastimes for people the same age, as it is for today's university students. For some ground personnel the war was an adventure; like the aircrew, they worked and played hard, and bent the rules: 'The youngsters today they wouldn't believe this old lady went doing these things. ... [We'd] never ask for a late pass. You'd just go out, and when we came back from dances we used to walk on the grass behind the guardroom and then ... carry on cycling. ... We weren't even supposed to be out.'

Ground personnel were vital to Bomber Command. Without them the aircraft could not fly. As the RAF expanded, thousands of men and women were recruited and trained in a variety of roles. They took pride in their work and enjoyed themselves while doing it, but then, as now, they received little recognition or thanks for their role. Written by an electrician stationed at RAF Tempsford in 1942, the following poem sums up the feelings of RAF ground personnel.

'Ode to the Erk'

Wherever you walk, you will hear people talk,
Of the men who go up in the air.
Of the dare-devil way, they go into the fray;
Facing death without turning a hair.
They'll raise a cheer and buy lots of beer,
For a pilot who's home on leave;
But they don't give a jigger
For a flight mech' or rigger
With nothing but 'props' on his sleeve.
They just say 'Nice day' and then turn away,
With never a mention of praise.
And the poor bloody 'erk' who does all the work;
Just orders his own beer And pays!
They've never been told of the hours in the cold
That he spends sealing Germany's fate.
How he works on a kite, till all hours of the night;
And then turns up next morning at eight.
He gets no rake-off for working till take-off;
Or helping the aircrew prepare;
But whenever there's trouble, it's 'Quick at the double';
The man on the ground must be there.
Each flying crew could tell it to you;
They know what this man's really worth.
They know he's part of the RAF's heart,
Even though he stays close to earth.
He doesn't want glory, but please tell his story;
Spread a little of his fame around.
He's one of the few so give him his due;
Three cheers for the man on the ground.

LOST AND FOUND

Arthur Perks (centre back) and members of No. 1 Missing Research and Enquiry Unit, out on the road, May 1947.

As a young man, Arthur Perks had always wanted to fly. He finally got his chance in 1944 at Elementary Flying Training School, but failed to master the vagaries of the DH82A Tiger Moth to his instructor's satisfaction. After almost fifteen hours of dual instruction, he was finally scrubbed and remustered as a navigator, training in South Africa before eventually being awarded his aircrew brevet in April 1945, three weeks before VE-Day.

Returning to the UK, Arthur was asked to attend an interview at the Air Ministry in Kingsway, and learned about a new organisation known as P4 (Cas) under the command of Group Captain Jimmy Hawkins. He was posted to No. 1 Missing Research and Enquiry Service (MRES) in Le Mans, later transferring to Chantilly. His commanding officer was Wing Commander Mike Shaw, a survivor of more than 100 bombing, minelaying and torpedo-dropping operations of the Second World War. Among his contemporaries was Frank Dolling, DFM, shot down in 1943. Dolling would later become Chairman of Barclays Bank International.

The MRES had been established to trace the whereabouts of the 42,000 RAF personnel who were listed as 'missing, believed killed', a task for which Arthur and his colleagues had no special training, neither did they have the advantage of modern forensic science. Identities could sometimes only be established by a scrap of uniform or the remnants of a smashed service watch, and often by a process of elimination. It could be a thankless and occasionally distressing task. In the villages and towns, local residents could be overwhelmingly helpful, or openly hostile, especially as the searches spread into a defeated Germany.

Despite the challenges, Arthur, and dozens like him, stuck relentlessly to the task over many years, identifying aircraft and their crews. By the time the Service was finally wound up, the MRES had managed to account for more than two-thirds of all those 'missing', and ensured that the airmen were given a formal burial within a Commonwealth War Graves Commission plot. It was a remarkable achievement, and brought peace of mind to thousands of relatives whose loved ones were no longer lost, but found.

FAILED TO RETURN

Wing Commander Mike Shaw (foreground) and his team helped bring peace of mind to hundreds of families of missing airmen.

Left The Memorial Spire. On the day the Spire was erected by Lindum Construction on 10 May 2015 the IBCC site was covered with poppies.

The Memorial Spire

The Spire is the centrepiece of the IBCC and the most recognisable element of the whole project. It is positioned on one of Lincoln's two hills, visible from miles around and in direct sight of Lincoln Cathedral, which is on top of the second hill. It is a lasting memorial to those who gave their lives. Echoing the cathedral's towers is particularly poignant, as the spires and churches around the county were used as sighting points by many of the crews as they left for their operations and for those lucky enough to return. More significantly, for many it was their last sight of their homeland before they gave their lives.

The generous gift of £600,000 from an anonymous donor in early 2015 allowed the construction and erection of the Spire. It is made up of two 'wings'. The larger is 102-feet (31.09 metres) high – the wingspan of the Avro Lancaster, which served the crews so well over the war years. The smaller 'wing' is 78.7-feet high – the wingspan of the Wellington bomber – another of the 'workhorses' of Bomber Command. At its base it is 16-feet (5 metres) wide – the width of a Lancaster wing. It was constructed in two parts, weighs some 73 tons, and is the UK's tallest war memorial.The material chosen for construction was Corten steel – or weathering steel, which has been developed to reduce the need for painting, as it forms a rust-like appearance over a number of years. Its vandal-resistant properties were, sadly, just as important a consideration, as stone is not an easy material from which to remove paint should vandalism occur.

The Spire made its journey from the manufacturers to the site on Sunday 10 May 2015 and arrived with a police escort at Canwick Hill. It took just eight hours to erect, with the official unveiling taking place in October 2015, by Lord Howe, Minister of State for the Ministry of Defence. The event was attended by over 300 of the remaining Bomber Command veterans (the oldest of whom was 102), together with their families and dignitaries from across the world. A poignant note of course is that it was one of the largest gatherings of veterans ever held, and it was for many also to be the last.

The unveiling also saw the first playing of the Bomber Command March – commissioned by the IBCC team and composed by Tom Davoren. The guests were treated to a spectacular flypast by a Vulcan bomber, a Dakota, two GR4 Tornados and three

Hawks, flying in tribute to the brave individuals the Centre remembers.

The Spire was awarded a 2016 Structural Steel Design Award in recognition of its high standard of architectural design, cost-effectiveness, aesthetics and innovation.

The Memorial Walls

The Spire is surrounded by the memorial walls. These are the cornerstone of the project as they contain the names of each of the individuals who bravely gave their lives for their country. From the very beginnings of the enterprise, it has been about the individuals rather than Bomber Command itself. It would have been easy to erect a memorial to Bomber Command as a group, and indeed the Bomber Command Memorial in London has done just this. But fitting though the London memorial is, it remembers the heroic efforts of the people as a group, rather than the individuals who made up that group. The 270 4mm-thick panels which make up the twenty-three walls are, like the Spire, manufactured from Corten steel, and will be weathered by the elements over time. The names themselves come from the losses database, and it is the only place in the world where all the deaths are commemorated.

The names, which are made up of around three-quarters of a million characters were laser cut and are arranged in alphabetical order. Where there are two or more people with the same name (a common occurrence), their name is annotated with the last digits of their service number, so that relatives can identify their loved ones. The

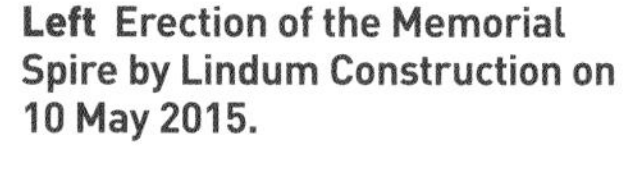

Left Erection of the Memorial Spire by Lindum Construction on 10 May 2015.

Opposite Remembering a father who is on the Wall of Names. The picture was taken at the Spire unveiling ceremony on 2 October 2015.

CLAYTON WR •
CLEASBY FJ • CLEAVER NC •
CLEGG J • CLEGG R •
CLEMENT C • CLEMENT CTL •
CLEMENTS AHT •
CLEMENTS HH •
CLEMENTS RG •
CLETHEROE C •
CLEVEY WA • CLIBURN EP •
CLIFTON AJ •
CLINKARD DCG • CLIPSON AL •
CLODE GE • CLOSE DE •
CLOUGH K • CLOUGH WP •
CLOUTING A • CLOVER AJ •
CLUFF RF • CLULOW GF •
CLYDE W • COAD AJ •
COAKES AE • COAKLEY GH •
COATES J • COATES JA •
COATES R 125 • COATES RJ •
COBB RE • COBB TA • COBB W •
COBERN CA • COCHRAN DH •
COCHRANE SG •
COCKBURN JT •
COCKERHAM L • COCKIE AM •

panels are numbered so that names can be found easily, and designed so that poppies can be placed as a mark of remembrance.

Importantly, no rank or medals are indicated on the walls as the project team feel that they all made the ultimate sacrifice and should be honoured as having done so. In addition of course, so many of those who perished were ground grew or died during training, and so did not get the opportunity to gain a rank or a position.

The Peace Gardens

The Centre has two peace gardens, set in around ten acres where visitors can quietly reflect on their thoughts.

The first, closest to the Spire, is the Lincolnshire Peace Garden. As Lincolnshire became quickly known as 'Bomber County', it was felt that this should be recognised at the Centre.

The Lincolnshire Peace Garden takes this further and contains twenty-seven native trees planted using coordinates to represent the airfields in Lincolnshire from where the aircraft operated. Interpretation in both physical and digital forms explains about each of the airfields and those who flew from them.

The International Peace Garden pays homage to the sixty-two nations who took part in Bomber Command and represent the five continents from where the individuals originated. It aims to further highlight the often untold story of the many nations who helped in the efforts and houses a series of beds using plants native to each continent.

The Chadwick Centre

The Chadwick Centre is the first port of call for many visitors to the site. It is named after Roy Chadwick who designed the iconic Lancaster that played such a large part in the lives of those who served in Bomber Command.

The centre houses the main visitor facilities such as ticketing, cafe and shop. Importantly, it is the introduction to the Centre and what it represents. As visitors engage with the exhibition they are introduced to the stories and most importantly the people who served in Bomber Command and who the centre remembers. The narrative begins with explanations about the Second World War – how it started, who was involved and how Bomber Command came into existence to help with the war effort.

The exhibition focuses very much on bringing to life the digital archive and losses database, with plenty of opportunity being given for visitors to find out more about their relatives, or others who were involved.

One of the main features follows twenty-four hours in the life of a Bomber Command raid, from initial briefings to the operation and follow-up activities. Films of actors the same age as the air- and ground crew deliver first-hand testimonies for visitors, who will understand the people behind the figures – one of the key aims of the project.

There is no hiding from the fact that Bomber Command is to many classed as 'difficult heritage' because of the consequences of their actions. However, the team were always clear that all accounts would be told in a human way that emphasises the challenges and difficulties on all sides – those who flew the aircraft on

Right The ground breaking for the Chadwick Centre on 3 October 2016, with Robert Carter (CEO of RG Carter Construction), Tony Worth CVO, Charles Clarke OBE, and Air Marshal Stuart Atha CB, DSO, ADC.

Below right The IBCC Chadwick Centre, named after Roy Chadwick who designed the iconic Avro Lancaster.

arduous missions as well as those who suffered as a result of their actions.

The story of Bomber Command is told by an 'orchestra of voices' using the interviews with veterans, both air- and ground crew, and support staff from around the world. Importantly, and additionally, there will be accounts of survivors of the Allied bombing campaign, those involved in the resistance movement and people affected by the influx of thousands of Service personnel into their lives. In each case they are the records of real people who were there, and who experienced the great upheaval the war bought upon their lives.

There are already many great stories of visitors coming to the centre, and discovering information they never knew about their relatives – such as how their air-craft was lost, the circumstances under which they were flying and the crews they were flying with. To have enriched these lives in this way makes the hard work so worthwhile.

Educating Future Generations

A critical strand of the Centre, and indeed the project, is that of education. The younger generations now are the first ones who will grow up with little or no direct contact with anyone who experienced the troubles of war at first hand. But it's so vital that this turbulent and life-changing period of time is not forgotten and so educating younger generations and those to come is of paramount importance to the team.

The Education Centre within the Chadwick Centre will hold purpose-written materials closely linked with the

National Curriculum, through which the stories will be told. Again, the focus will be on the people involved, but with an appreciation to the wider issues, such as why the war happened and what impact it had on children and young adults. In addition, close liaison with schools will enable tours around the centre to be both informative and memorable. Pre-visit worksheets, guided visits and tours of the memorial and follow-on task will help to keep the sacrifices made in the minds of the pupils, at the same time as informing them about a significant time in the history of the country in which they live.

A Fitting and Lasting Memorial

When the International Bomber Command Centre opened its doors in January 2018 it was the culmination of years of hard work by hundreds of dedicated professionals and volunteers.

On a practical scale it was many things – a complex and challenging building project, a data management exercise and a prime example of community engagement. On a wider note it is a case study in collaboration between academia, the corporate world, government bodies, the armed forces, masses of volunteers and local communities.

It had been quite literally a race against time and over that time the project team became very real friends with the veterans who provided their stories and total support so that their colleagues could be remembered. Such a bond between the veterans and the team made their passing so much harder, but it reinforced the team's determination to complete the Centre, and just as importantly to make sure

Left A key focus of the IBCC has been to educate future generations. The picture was taken in November 2015 on the first of what has become an annual remembrance service for primary school children.

Right Bomber Command veteran James Flowers, who flew as an air gunner with No. 50 Squadron, at the April 2016 Queen's birthday beacon lighting.

it was done right.

It provides a lasting memorial from the many to the few, who gave so much so that we can live in a free world. It also provides the only place in the world where the losses are recorded in such detail, and where the personal accounts behind those who lost their lives can be told. So what does the future hold for the International Bomber Command Centre?

The team were clear at the outset that they wanted to construct a fitting and lasting memorial to those who lost their lives, and no one seeing the Memorial Spire proudly reaching to the sky, surrounded by the walls containing the names of 58,000 individuals, can doubt that this has been done.

They were just as determined that the losses database would be a resource to work from, and already the data is being enriched by the relatives of those who are listed in the archive with narratives of their lives and families.

The team also hoped that the Centre would help to educate adults and children alike in the human story behind Bomber Command and those who gave their lives, and the feedback received from visitors, schools and children so far are testament to this being achieved. From the initial thoughts of Tony Worth in Lincoln Cathedral, a great project evolved, which brought people together for a common and worthwhile cause. It had been an exercise in collaboration and compassion, motivation and logistics, but above all it showed how people working together can achieve great things, and now it is complete the lives of those who gave so much will be always remembered.

ABOUT THE AUTHORS

Steve Darlow
Author of The Aircrew Story
Steve is a Bomber Command historian, established military aviation author (with twenty books to his name), magazine contributor, and television documentary consultant and contributor. In 2009 Steve founded Fighting High Publishing, which seeks to publish books focusing on human endeavour in military situations.

Mark Dodds
Author of The International Bomber Command Centre
Mark is Marketing Manager for a large regional law firm. Combining copywriting with an interest in Second World War history, he sits on the IBCC Development Panel, where he helps to raise the profile of this important project.

Dr Dan Ellin
Author of RAF Bomber Command Ground Personnel
Dr Ellin is the IBCC's Archive and Exhibition Curator. He is a graduate of Lincoln University, and studied Bomber Command ground personnel for his PhD from the University of Warwick.

Sean Feast
Author of RAF Bomber Command Failed to Return
Sean works in PR and advertising and is an acclaimed Bomber Command historian and author, including numerous volumes of Fighting High's 'Failed to Return' series, and 'Thunder Bird in Bomber Command'.

Dr Robert Owen
Author of RAF Bomber Command at War
Dr Owen is an aviation historian and the Official Historian of the No. 617 Squadron Association. His interest in the Squadron extends over 40 years and he has authored and co-authored numerous books, including 'Henry Maudslay – Dambuster', 'Dam Busters – Failed to Return', 'V-Weapons Failed to Return', and 'Battle of Berlin – Failed to Return'. He has assisted authors of numerous authoritative titles on Bomber Command together with a number of TV documentaries. He has also contributed to publications for the RAF Museum including 'Breaching the German Dams'. He is a contributor to magazines including Flypast, Aeroplane and Britain at War.

SOURCES AND ACKNOWLEDGEMENTS

The publisher wishes to express their sincere thanks to Nicky Barr, Sue Taylor and their team of IBCC volunteers for their support in putting this book together. Also, a special thanks to Paul Mellor for providing his photography expertise.

Three Cheers for the Men and Women on the Ground.
Winifred Barker, ABarkerWG160912, IBCC Digital Archive. Basil Goldstraw, AGoldstrawBJ160827, IBCC Digital Archive. John Bagg, ABaggJG160902, IBCC Digital Archive. Margaret Young, AYoungM150515, IBCC Digital Archive. Charles Bland, ABlandC150817, IBCC Digital Archive. Private Papers of Mrs E.M. Kup, Documents 507, Imperial War Museum. Private Papers of Mrs S. Watts, Documents 5652, Imperial War Museum. Ted Mawdsley, AMawdsleyT150502, IBCC Digital Archive. Kenneth Locke Brown, ALockeBrownK1150706, IBCC Digital Archive. Margaret Bailey, ABaileyLMM151006, IBCC Digital Archive. E. Howell, AC96/5. 'Reminiscences by former members of Bomber Command gathered in research for the writing of The Hardest Victory', RAF Museum, Hendon. Howell, E. The OR's Story: Bomber Command Other Ranks in World War Two (Swindon, NPG, 1998), p. 119. Dennis Brett, ABrettD150522, IBCC Digital Archive. Gladys Gildersleve, AGildersleveG160905, IBCC Digital Archive. Anon. 'Reminiscences by former members of Bomber Command gathered in research for the writing of The Hardest Victory', RAF Museum, Hendon. J. Johnstone, AC96/5. 'Reminiscences by former members of Bomber Command gathered in research for the writing of The Hardest Victory', RAF Museum, Hendon. F.L. Warner. AC96/5. 'Reminiscences by former members of Bomber Command gathered in research for the writing of The Hardest Victory', RAF Museum, Hendon. Air Ministry and Ministry of Defence: Operations Record Books, Royal Air Force Stations. Waterbeach, Appendices, AIR 28/891, The National Archives. Private Papers of E.M. Kup, Documents 507, Imperial War Museum. Joyce Bell, ABellJ151127, IBCC Digital Archive. E. Sykes, 'Three Cheers for the Men on the Ground', The Wickenby Register Newsletter (1988), No. 26, p. 8. RAF Wickenby Memorial Collection. The comedian and actor of the same name was also in the RAF but served with a mobile signals unit in Europe.

The Aircrew Story
Author interviews and correspondence with the following veterans: Jo Lancaster DFC, George Dunn DFC L d'H, Dave Fellowes L d'H, Gerry Norwood, Hal Gardner, Steve Bethell, Henry Wagner, Gordon Mellor, Jack Smith, Roy Smith, Jack Pragnell. Interviews held in the IBCC Digital Archive: Ken Johnson (AJohnsonK150603-Transcription (Bracknall)), Alan McDonald (AMcDonaldEA150713-Transcription (Williams)). Mike Connock (No.50 & No.61 Squadrons Association). Thank you to Janet Mooney, Dan Smith, Deborah Evans, Madeleine Taylor, and Rob Long for their assistance with the veteran photoshoot.

Failed to Return
Bertie Booth and his friends Harrie (Hein) Jansen and Wim Slangen; the late John Ward; fellow author Dr Steve Bond; Stuart Green; Peter Wheeler QSM of the New Zealand Bomber Command Association; Les Owen of the former Little Staughton Pathfinder Association; Peter Coulter of the 550 Squadron Association; Caroline Feast; and Maureen Olley.

PHOTOGRAPHY

Unless indicated below the image, all modern photographs are courtesy of the International Bomber Command Centre and all Second World War photographs are courtesy of the International Bomber Command Centre Digital Archive.

Paul Mellor 8–9, 36, 53, 54, 66, 67, 70, 75, 78, 86, 89, 91, 92, 95
Air Historical Branch 22, 31
Steve Darlow 21, 36, 63, 64 (picture 1), 80, 105
Roy Smith 65
George Dunn 67 (right)
Gerry Norwood 68 (right), 77
Steve Bethell 71, 73
Hal Gardner 76
Jack Pragnell 84
Ken Johnson 89
Declan O'Flanagan 109 (top left)
Courtesy of Dr Steve Bond 20
Courtesy of the late John Ward 36,37
Via Peter Wheeler QSM and the NZBCA 52
Via Stuart Green 62
Courtesy of Judy Bradley and Elva Scholz via Bertie Booth 79
Courtesy of Les Owen 96
Via Peter Coulter of the 550 Squadron Association 104
Courtesy of Caroline Feast 114
Courtesy of Maureen Olley 115

INDEX